TEXTILE EFFLUENT TREATMENT METHODS
(A Current Focus on Research and Developments)

TEXTILE EFFLUENT TREATMENT METHODS
(A Current Focus on Research and Developments)

Editor

Dr. O. L. Shanmugasundaram

WOODHEAD PUBLISHING INDIA PVT LTD

New Delhi

Published by Woodhead Publishing India Pvt. Ltd.

Woodhead Publishing India Pvt. Ltd.,
303, Vardaan House, 7/28, Ansari Road,
Daryaganj, New Delhi - 110002, India
www.woodheadpublishingindia.com

First published 2020, Woodhead Publishing India Pvt. Ltd.
© Woodhead Publishing India Pvt. Ltd., 2020

Woodhead Publishing India Pvt. Ltd. ISBN: 978-93-88320-31-3
Woodhead Publishing India Pvt. Ltd. e-ISBN: 978-93-88320-32-0

Typeset by Cognition Technology
Printed and bound in India by Replika Press Pvt. Ltd.

Contents

Preface

Water is one of the utmost important natural resources and also one of the basic needs for living things to stay alive in the world. Water is mainly used for drinking, irrigation and various processes in industries. In the current scenario, industrial sectors are consuming huge volume of water for processes and discharging the same as wastewater either into the land or water stream. This causes environmental pollution and water scarcity. Wastewater contains many contaminants which are harmful to the living things. Therefore, the contaminated water has to be treated and may be reused reuse for many purposes.

An Effluent treatment plant is a process design for treating the industrial waste water for its reuse or safe disposal to the environment. The existing wastewater treatment methods involve a laborious process, not economically viable and are difficult to manage solid waste disposal; therefore, industries are not interested in taking up wastewater treatment methods.

Technologists are finding innovative wastewater treatment techniques to reduce capital and operation cost, energy and water consumption, solid waste generation, floor space area and skilled labour. These techniques safeguard the environment against pollution and contribute to sustainable development and to meet the standards of pollution control board.

New techniques are being developed by Scientists and research scholars to treat industrial effluents either to meet standards of drinking water or irrigation water or to be reused for industrial purposes. This book provides an information on current research and developments in effluent treatment techniques and their limitations and scope for further advancements. The academicians and research scholars from leading institutions / universities in India and abroad have contributed their research works or review work as a chapter in this textbook. This book deals with explicit features in emerging wastewater treatment techniques for effective removal of pollutants. A sustainable combined wastewater treatment technique is capable of removing colour, metal and other pollutants, has immense potential for future industrial applications.

The chapters in this book have been arranged in three areas which includes nanotechnology approach, green chemistry method and sustainable approaches. An attempt has been made to solve practical difficulties confronting the textile industries. This book will be useful for academicians, researchers, environmental and textile engineers, pollution control board, policy makers, stake holders and government and public organizations.

At this juncture, I would like to express my sincere thanks to all authors and also grateful to our contemporaries who helped directly and indirectly for the successful completion of this book. I honestly acknowledge the reviewers for reviewing the chapters and also express my gratitude to my family members for extending their support during the preparation of this book.

1

Introduction and importance of wastewater treatments

Shanmugasundaram O. Lakshmanan[a]* **and**
Sujatha Karuppiah[b]

[a]*Department of Textile Technology, K. S. Rangasamy College of Technology,*
Tiruchengode, Tamil Nadu, India
[b]*Department of Physics, Vellalar College for Women, Erode, Tamil Nadu, India*
Corresponding Author. E-mail: mailols@yahoo.com

Abstract: Wastewater treatment is a process to remove pollutants (contaminants) and to improve the quality and purity of water for reuse or safe disposal to the environment. In many parts of the world, including India is currently suffering from water scarcity. Freshwater and saltwater are polluted by discharging of untreated wastewater from industries. The untreated or inadequately treated wastewater causes health and diseases problems to living organisms, minimum environmental degradation, and destruction of marine life. The polluted water has a serious issue like use of water for drinking, fishing, agriculture, and transportation and also adverse effect on all living organisms. This chapter offered the importance of wastewater treatments, major pollutants present in textile effluents and their adverse effect on environment, and emerging wastewater treatment techniques.

Keywords: wastewater; reuse; recycle; treatment techniques.

1.1.   Introduction

A wide variety of impurities are present in textile wastewater, which contain fibers, dyes, and chemicals. Various type of dyes are used by textile industries such as direct, reactive, vat, disperse, acid, and basic dyes. The pollutants such as salts, surfactants, dissolved solids, dyes, chlorine, and other auxiliary chemicals are present in textile wastewater. These pollutants are toxic, carcinogenic, and mutagenic and have to be removed before discharging into water stream/environment [1].

Textile industries are generating huge volume of effluent from various processes such as desizing, scouring, bleaching, dyeing, printing, and finishing. The quality of fresh water is deteriorating due to discharge of textile effluent into water streams [23,59]. Correia et al. [15] stated that toxic pollutants (such as non-biodegradable, pesticides, and highly colored dyes) are present in industrial effluents. Nigam et al. [44] stated that textile effluent containing very low quantity of dyes.

A large amount of fresh water, inorganic compounds, organic chemicals, and polymers are used in wet processing of textiles [9,28,40]. The treated water has to meet the pollution control board standards or legal requirements for discharge of wastewater into water stream. Lots of research works have been reported that membrane filtration techniques and adsorbents are most widely used methods to treat effluents for removal of heavy metals [10,18,27]. Recently, works have been carried out for the development of new technologies and techniques to decrease the generation of quantity of effluent and to treat wastewater effectively for reuse purpose.

1.2.   Complications of wastewater on environment

Robinson et al. [49] stated that seven lakh tons of dyes are produced annually. Coloring matter has to be removed otherwise it creates a visual problem and contribute to the load of BOD (biochemical oxygen demand). Koltuniewicz and Drioli [31] found that more than 2000 parts per million (ppm) salts present in reactive dye effluent, which cause aquatic life damage and soil infertility.

The adsorption, membrane filtration, ion-exchange, coagulation, and flocculation techniques are used to remove toxic metals (for example Cu, Cd, Ni, Zn, and Cr) present in industrial effluents [19]. These metals are toxic to living organisms and have adverse effects on human and ecology. Researchers found the chemical composition of textile effluent and their adverse effect on ecology [9,39]. A huge quantity of heavily metals such as Cu, Cd, Cr, Ni, Pb, and Zn are discharged into environment which causes health disorders [8]. Heavy metals present in water/wastewater cause health problems such as kidney damage, cancer, brain damage, organ failure, etc.

The textile industries are using carcinogenic azo dyes which are toxic and resistant to microbial degradation and lead to environmental issues. Shore [50] found that basic and direct dyes are highly toxic among other class of dyes. The fly-ash has been used as an adsorbent for removal of metal complex dyes in textile effluent (Nasser and Elgeaendi, 1991) [61]. Basava Rao and Ram Mohan Rao [11] found that 90–100% dye removal is possible and the results were followed by Langmuir model. The limitation/drawback of this process is disposal of treated fly-ash, which is a major problem.

1.3.   Emerging wastewater treatment techniques

1.3.1   Coloring removal techniques

Many authors have reported in the literature on color removal techniques and its mechanisms. The physical dye separation, decolorization by adsorption, and breakdown of dyes are the techniques available for removal of color in effluent. Coagulation technique is effective for removal of disperse and sulfur dyes in textile effluent. However, huge quantity of sludge is generated during the processes, and disposal of sludge is a major problem in industries.

Microbial decolorization techniques are effective to degrade certain dyes but their limitations are time consuming, require large floor space area, and microorganisms are sensitive towards certain chemicals. Hao et al. [21] found that microbial technique is not suitable for certain class of dyes due to their toxicity. Wet air oxidation processes are useful to treat effluents containing toxic substances/chemicals. However, their application is limited to high installation and operating cost.

Ozone technique is more effective for removal of color present in effluent due to high oxidation potential of ozone. However, their application is limited as it requires an ozone destruction unit. Ghoreishi and Haghighi [20] conducted an investigation on treatment of textile effluent by combining chemical catalytic reaction and biological oxidation method for removal of color (75–85%) and decrease in BOD (90%), COD (83%), and TDS (95%). Few scientists used fly-ash, saw dust, peat, and bagasse pitch for removal of color present in textile wastewater. Pagga and Taeger [46] developed a new method using activated sludge for adsorption of dyestuff. In 1976, Poots and his group used natural adsorbents for removal of color in effluent [62]. Hu [25] examined the efficiency of color removal from reactive dye effluent using different bacterial genera.

Membrane filtration technique is capable of removing dyes without generation of sludge but their limitations are lower productivity, high cost of membrane and maintenance, membrane fouling, and disposal of concentrates. Aouni and his research team used ultrafiltration and nanofiltration techniques to treat reactive dye effluent and obtained good conductivity rates (80%), high COD retentions (80–90%), and high color retention (more than 90%) [7].

Many scientists found that complete removal of dye and pollutants are achieved by combination of two or more effluent treatment techniques [35,37]. Nanofiltration technique is found to be suitable to treat textile effluent for water reuse in industry [52]. An extensive research works has been carried for the recovery and reuse of wastewater using membrane techniques such as microfiltration [60], ultrafiltration [2,17,38,60], nanofiltration [4,6,13,54,60], and reverse osmosis [5,17,43,60]).

1.3.2   Metal removal techniques

Eccles [16] stated that heavy metals can be removed by ion exchange and electrochemical processes. However, their limitations are incomplete removal and high energy consumption. Leung et al. [34] identified low adsorbents for the removal of metal in wastewater. Kurniawan et al. [33] used low cost adsorbents (agriculture wastes, bio mass, and polymeric materials) for the removal of Cr(IV) in industrial effluent. Photocatalytic method is found to be capable of removing pollutants in water [51].

Wang et al. [56] used chemical precipitation method for removal of heavy metals present in wastewater and demonstrated the mechanism of removal of heavy metals in effluents. Ion exchange and electrolytic recovery methods are used to remove heavy metal from effluent. These methods have some limitations like highly sensitive to pH and corrosion. New adsorbents, modified natural materials, industrial by-products, modified agriculture and biological wastes, and modified biopolymers are used as adsorbents for the removal of heavy metals from wastewater [10].

Rether and Schuster [48] reported that polymer based/supported ultrafiltration process is a promising technique for effective removal of heavy metals from effluents. The cation-exchange membrane perfluorosulfonic Nafion 117 is found to be suitable for removal of Co(II) (90%) and Ni(II) (69%) by Tzanetakis et al. [53]. Barakat [10] published a review paper on new techniques for removal of heavy metal from industrial effluents. Moraes and their team members combined photocatalytic and ozonation processes for complete degradation of reactive dyes and reduction of toxicity from textile effluent [42].

1.3.3   Other pollutants removal techniques

Some interesting results were reported by Robinson et al. [49] on current effluent treatment technologies for effective removal of pollutants in industrial wastewater. Ozone technique is capable of removing color, COD, BOD, and degrading phenols, pesticides, chlorinated and aromatic hydrocarbons present in industrial effluent. Moreover, the treated wastewater is suitable for discharging into water streams [36,57]. Electrochemical destruction technique is found to be efficient for degradation of pollutants and removal of color [45,47]. There have been many literature surveys which shows that research works are being conducted to reduce the use of water, effectively treat the effluent, and minimize sludge production [14, 22, 24, 26, 41].

Electrocoagulation is an environment-friendly technique to treat textile effluent with minimum sludge generation and eliminate harmful chemicals and the major limitation is high operating cost. Many researchers [3, 12, 29,

32, 55, 58] used electrocoagulation technique to treat textile wastewater. Khandegar and Saroha [30] reported the advantages and limitations of electrocoagulation technique for the treatment of textile effluent and clearly stated the benefits such as versatility, safety, easy operation, less amount of sludge generation, etc. Many researchers found that combination of electrocoagulation technique with other methods remove major pollutants present in textile effluent.

1.4   Conclusion

Protection of environment and human health is utmost important than the profitability of an industry. Thus, the scientists and research scholars are developing new techniques to treat industrial effluents either to meet the standards of drinking water or irrigation water or reuse water quality for industrial purpose. In recent years, pollution control board is proclaiming many stringent norms for discharging of effluent into water stream. Henceforth, textile industries have to treat the effluent in order to meet standards. Moreover, biotechnologists, civil engineers, and chemical technologists have to find innovative wastewater treatment techniques to reduce capital and operation cost, energy and water consumption, solid waste generation, floor space area, and skilled labor. The new techniques may safeguard the environment against pollution and contribute for sustainable development.

References

1. Adhoum, N.; Monser, L.; Bellakhal, N.; Belgaied, J. Treatment of electroplating wastewater containing Cu^{2b}, Zn^{2b} and Cr (VI) by electrocoagulation. *J. Hazard. Mater.* **2004**, *B112*(3), 207–213.

2. Ahmad, A. L.; Puasa, S. W. Reactive dyes decolourization from an aqueous solution by combined coagulation/micellar-enhanced ultrafiltration process. *Chem. Eng. J.* **2007**, *132*, 257–265.

3. Aji, B. A.; Yavuz, Y.; Koparal, A. S. Electrocoagulation of heavy metals containing model wastewater using monopolar iron electrodes. *Sep. Purif. Technol.* **2012**, *86*, 248–254.

4. Akbari, A.; Desclaux, S.; Rouch, J. C.; Remigy, J. C. Application of nanofiltration hollow fibre membranes, developed by photografting, to treatment of anionic dye solutions. *J. Membr. Sci.* **2007**, *297*, 243–252.

5. Allegre, C.; Moulin, P.; Maisseu, M.; Charbit, F. Savings and re-use of salts and water present in dye house effluents. *Desalination.* **2004**, *162*, 13–22.

6. Aouni, A.; Fersi, C.; Ben Sik Ali, M.; Dhahbi, M. Treatment of textile wastewater by a hybrid electrocoagulation/nanofiltration process. *J. Hazard. Mater.* **2009**, *168*, 868–874.

7. Aouni, A.; Fersi, C.; Cuartas-Uribe, B.; et al. Reactive dyes rejection and textile effluent treatment study using ultrafiltration and nanofiltration processes. *Desalination.* **2012**, *297*, 87–96.

8. Babel, S.; Kurniawan, T. A. Cr(VI) removal from synthetic wastewater using coconut shell charcoal and commercial activated carbon modified with oxidizing agents and/or chitosan. *Chemosphere.* **2004**, *54*(7), 951–967.

9. Banat, I. M.; Nigam, P.; Singh, D.; Marchant, R. Microbial decolorization of textile-dye containing effluents: A review. *Bioresour. Technol.* **1996**, *58*, 217–227.

10. Barakat, M. A. New trends in removing heavy metals from industrial wastewater. *Arab. J. Chem.* **2011**, *4*, 361–377.

11. Basava Rao, V. V.; Ram Mohan Rao, S. Adsorption studies on treatment of textile dyeing industrial effluent by flyash. *Chem. Eng. J.* **2006**, *116*, 77–84.

12. Bektas, N.; Akbulut, H.; Inan, H.; Dimoglo, A. Removal of phosphate from aqueous solutions by electro-coagulation. *J. Hazard. Mater. B.* **2004**, *106*, 101–105.

13. Bes-Piá, A.; Iborra-Clar, M. I.; Iborra-Clar, A.; Mendoza-Roca, J. A.; Cuartas-Uribe, B.; Alcaina-Miranda, M. I. Nanofiltration of textile industry wastewater using a physicochemical process as a pre-treatment. *Desalination.* **2005**, *178*, 343–349.

14. Cheug, P. K.; Chen, K. C. Investigation of color removal technology. In *Proceedings of the 16th Conference on Wastewater Treatment Technology*, China; **1991**, 613–619.

15. Correia, V. M.; Stephenson, T.; Judd, S. J. Characterisation of textile watsewaters—A review. *Environ. Technol.* **1994**, *15*, 917–929.

16. Eccles, H. Treatment of metal-contaminated wastes: Why select a biological process? *Trends Biotechnol.* **1999**, *17*, 462–465.

17. Fersi, C.; Dhahbi, M. Treatment of textile plant effluent by ultrafiltration and/or nanofiltration for water reuse. *Desalination,* **2008**, *222*, 263–271.

18. Fersi, C.; Gzara, L.; Dhahbi, M. Treatment of textile effluents by membrane technologies. *Desalination.* **2005**, *185*, 1825–1835.

19. Fu, F.; Wang, Q. Removal of heavy metal ions from wastewaters: A review. *J. Environ. Manage.* **2011**, *92*(3), 407–418.

20. Ghoreishi, S. M.; Haghighi, R. Chemical catalytic reaction and biological oxidation for treatment of non-biodegradable textile effluent. *Chem. Eng. J.* **2003**, *95*, 163–169.

21. Hao, O. J.; Kim, H.; Chiang, P. C. Decolorization of wastewater. *Crit. Rev. Environ. Sci. Technol.* **2000**, *30*(4), 449–505.

22. Hitz, H. R.; Huber, W.; Reed, R. The adsorption of dyes on activated sludge. *J. Soc. Dyers Colorists.* **1978**, *94*, 71–76.

23. Hu, T. L. Separation of reactive dye aeromonas biomass. *Water Sci. Technol.* **1992**, *26*, 357–366.

24. Hu, T. L. Decolorization of reactive azo dyes by transformation of *Pseudomonas luteola*. *Bioresour. Technol.* **1994**, *49*, 47–51.

25. Hu, T. L. Removal of reactive dyes from aqueous solution by different bacteria genera. *Water Sci. Technol.* **1996**, *34*, 89–95.

26. Hu, T. L.; Ko, W. L. Adsorption of reactive dyes by biomass. In *Proceedings of the 17th Conference on Wastewater Treatment Technology*, China; **1992**, 05–116.

27. Juang, R. S.; Shiau, R. C. Metal removal from aqueous solutions using chitosan-enhanced membrane filtration. *J. Membr. Sci.* **2000**, *165*, 159–167.

28. Juang, R. S.; Tseng, R. L.; Wu, F. C.; Lin, S. J. Use of chitin and chitosan in lobster shell wastes for colour removal from aqueous solutions. *J. Environ. Sci. Health A.* **1996**, *31*, 325–338.

29. Khandegar, V.; Saroha, A. K. Electrocoagulation for the treatment of textile industry effluent: A review. *J. Environ. Manag.* **2013a**, *128*, 949–963.

30. Khandegar, V.; Saroha, A. K. Electrochemical treatment of effluent from small scale dyeing unit. *Indian Chem. Eng.* **2013b**, *55*(2), 1–9.

31. Koltuniewicz, A. B.; Drioli, E. *Membrane in Clean Technologies: Theory and Practice*; **2008**, Vol. 1. Wiley-VCH, Berlin.

32. Kumar, P. R.; Chaudhari, S.; Khilar, K. C.; Mahajan, S. P. Removal of arsenic from water by electrocoagulation. *Chemosphere.* **2004**, *55*(9), 1245–1252.

33. Kurniawan, T. A.; Chan, G. Y. S.; Lo, W. H.; Babel, S. Comparisons of low-cost adsorbents for treating wastewaters laden with heavy metals. *Sci. Total Environ.* **2005**, *366*(2–3), 409–426.

34. Leung, W. C.; Wong, M. F.; Chua, H.; Lo, W.; Leung, C. K. Removal and recovery of heavy metals by bacteria isolated from activated sludge treating industrial effluents and municipal wastewater. *Water Sci. Technol.* **2000**, *41*(12), 233–240.

35. Lin, S. H.; Chen, M. L. Treatment of textile wastewater by chemical methods for reuse. *Water Res.* **1997**, *31*, 868–876.

36. Lin, S. H.; Lin, C. M. Treatment of textile waste effluents by ozonation and chemical coagulation. *Water Res.* **1993**, *27*, 1743–1748.

37. Lin, S. H.; Peng, C. F. Continuous treatment of textile wastewater by combined coagulation, electrochemical oxidation and activated sludge. *Water Res.* **1996**, *30*, 587–592.

38. Majewska-Nowak, K. The effect of a polyelectrolyte on the efficiency of dye-surfactant solution treatment by ultrafiltration. *Desalination.* **2008**, *221*, 395–404.

39. Mishra, G.; Tripathy, M. A critical review of the treatments for decolourization of textile effluent. *Colourage.* **1993**, *40*, 35–38.

40. Mishra, G.; Tripathy, M. A critical review of the treatments for decolourization of textile effluent. *Colourage.* **1993**, *40*, 35–38.

41. Mittal, A. K.; Gupta, S. K. Biosorption of cationic dyes by dead macro fungus *Fomitopsis carnea*: Batch studies. *Water Sci. Technol.* **1996**, *34*, 81–87.

42. Moraes, S. G.; Freire, R. S.; Duran, N. Degradation and toxicity reduction of textile effluent by combined photocatalytic and ozonation processes. *Chemosphere.* **2000**, *40*, 369–373.

43. Nataraj, S. K.; Hosamani, K. M.; Aminabhavi, T. M. Nanofiltration and reverse osmosis thin film composite membrane module for the removal of dye and salts from the simulated mixtures. *Desalination.* **2009**, *249*, 12–17.

44. Nigam, P.; Armour, G.; Banat, I. M.; Singh, D.; Marchant, R. Physical removal of textile dyes and solid state fermentation of dye-adsorbed agricultural residues. *Bioresour. Technol.* **2000**, *72*, 219–226.

45. Ogutveren, U. B.; Kaparal, S. Colour removal from textile effluents by electrochemical destruction. *J. Environ. Sci. Health A.* **1994**, *29*, 1–16.

46. Pagga, U.; Taeger, K. Development of a method for adsorption of dyestuff on activated sludge. *Water Res.* **1994**, *28*, 1051–1057.

47. Pelegrini, R.; Peralto-Zamora, P.; de Andrade, A. R.; Reyers, J.; Duran, N. Electrochemically assisted photocatalytic degradation of reactive dyes. *Appl. Catal B-Environ.* **1999**, *22*, 83–90.

48. Rether, A.; Schuster, M. Selective separation and recovery of heavy metal ions using water-soluble N-benzoylthiourea modified PAMAM polymers. *React. Funct. Polym.* **2003**, *57*, 13–21.

49. Robinson, T.; McMullan, G.; Marchant, R.; Nigam, P. Remediation of dyes in textile effluent: A critical review on current treatment technologies with a proposed alternative. *Bioresour. Technol.* **2001**, *77*, 247–255.

50. Shore, J. Advances in direct dyes. *Indian J. Fib. Text. Res.* **1996**, *21*, 1–29.

51. Skubal, L. R.; Meshkov, N. K.; Rajh, T.; Thurnauer, M. Cadmium removal from water using thiolactic acid-modified titanium dioxide nanoparticles. *J. Photochem. Photobiol. A, Chem.* **2002**, *148*, 393.

52. Tang, C.; Chen, V. Nanofiltration of textile wastewater for water reuse. *Desalination.* **2002**, *143*, 11–20.

53. Tzanetakis, N.; Taama, W. M.; Scott, K.; Jachuck, R. J. J.; Slade, R. S.; Varcoe, J. Comparative performance of ion exchange membrane for electrodialysis of nickel and cobalt. *Sep. Purif. Technol.* **2003**, *30*, 113–127.

54. Van der Bruggen, B.; Braeken, L.; Vandecasteele, C. Flux decline in nanofiltration due to adsorption of organic compounds. *Sep. Purif. Technol.* **2002**, *29*, 23–31.

55. Verma, S. K.; Khandegar, V.; Saroha, A. K. Removal of chromium from electroplating industry effluent using electrocoagulation. *J. Hazard. Toxic. Radio. Waste.* **2013**, *17*(2), 146–152.

56. Wang, L. K.; Vaccari, D. A.; Li, Y.; Shammas, N. K. Chemical precipitation. In: Wang, L. K.; Hung, Y. T.; Shammas, N. K. (Eds.), *Physicochemical Treatment Processes*; **2004,** vol. 3. Humana Press, New Jersey, pp. 141–198.

57. Xu, Y.; Lebrun, R. E. Treatment of textile dye plant effluent by nanofiltration membrane. *Sep. Sci. Technol.* **1999**, *34*, 2501–2519.

58. Yuksel, E.; Eyvaz, M.; Gurbulak, E. Electrochemical treatment of colour index Reactive Orange 84 and textile wastewater by using stainless steel and iron electrodes. *Environ. Prog. Sust. Energy.* **2013**, *32*, 60–68.

59. Yu-Li Yeh, R.; Thomas, A. Color difference measurement and color removal from dye waste water using different adsorbents. *J. Chem. Technol. Biotechnol.* **1995**, *63*, 55–59.

60. Zaghbani, N.; Hafiane, A.; Dhahbi, M. Removal of Safranin T from wastewater using micellar enhanced ultrafiltration. *Desalination.* **2008**, *222*, 348–356.

61. M. Nassar, M. Elgeaendi, Comparative cost of color removal from textile effluents using natural adsorbents J. Chem. Technol. Biotechnol **1991**, *50*, 257–264.

62. Poots V., Mckay G. and Healy J., "The Removal of Acid 51ye from Effluent Using Natural Adsorbents", Wat. Res, **1976**, *10*, 1067-1074.

2

Nanocellulose for the removal of textile dyes from water

Deepu A. Gopakumar[a,b,*], Nandakumar[c], Ange Nzihou[a],
Nathalie Lyczko[a], Abdul Khalil, H.P.S[d], Sabu Thomas[b]

[a]*Université de Toulouse, Mines Albi, RAPSODEE CNRS UMR-5302,*
Campus Jarlard, F-81013 Albi cedex, 09, France
[b]*International and Inter University Centre for Nanoscience and Nanotechnology,*
Mahatma Gandhi University, Kottayam, Kerala 686560, India
[c]*Amrita School of Arts and Science, Amrita Viswa Vidyapeedam, Amrithapuri,*
Kollam, Kerala, India
[d]*School of Industrial Technology, Universiti Sains Malaysia, Penang, Malaysia*
*Corresponding author: deepu1789@gmail.com

Abstract: The usage of textile dyes has an adverse effect on all forms of life. It has been showed that the textile dyes together with enzymes chromium compounds and heavy metals make the textile effluent highly toxic. Natural dyestuffs need huge amount of water for dyeing. About 80% of the dyestuffs stay on the fabric, while the rest go down the drain. The textile dyeing and finishing industry has created a huge pollution problem as it is one of the most chemically intensive industries on earth. More than 3600 individual textile dyes are being manufactured by the Industry today. Since waste water is complex, the removal of textile dyes from water is a major challenge. Different methodologies like reverse osmosis, nanofiltration, and adsorption using activated materials have been employed for the removal of textile dye effluents from water. Among these techniques' adsorption has been found to be the most effective and cost-effective process for the removal of dyes from the waste water. In this context, nanocellulose has emerged as a cost-effective sustainable absorbent towards the textile dyes via electrostatic attraction. Nanocellulose's inherent nature like high aspect ratio, specific area, excellent strength, etc. made nanocellulose as a promising candidate for the adsorption of textile dyes from water. Moreover, nanocellulose has large number of surface hydroxyl groups which facilitates a variety of surface modification which resulted in the removal of the various pollutants from the water rather than dyes. This chapter is focused on the importance of nanocellulose in the water purification especially the removal of toxic textile dyes from water.

Keywords: Nanocellulose; adsorption; textile dye removal; water purification.

2.1   Introduction

Of the many resources that humans are intensely dependent on, water plays a great role. Water be it potable or not is of crucial usage in day to day activities. Starting up from basic need of eating and drinking, even to usage as a coolant in many industries, water is an indispensable part of daily routine. Not only humans but plants and animals also need water to survive. However, this resource is nowadays intensively used and polluted. Many industries use water bodies as a site for the dumping of their wastes. Also, they remain ignorant of the effect that such dumping can cause on the lives of people as well as other living things. Heavy metals, organic compounds like dyes, and non-biodegradables like plastic, etc. adversely affect the life of living things. Among all others dyes mainly synthetic dyes are the dogged pollutants which can cause effects like low chemical oxygen demand (COD) to even mutations via interaction with DNA. Synthetic dyes are produced annually about 700,000 tons [1]. These are widely used in industries like leather, textiles, paper, etc. The wide use means that the chance for contamination is higher and thus dye removal is to be considered with extreme urgency. If left untreated dyes cause above mentioned problems like high COD and further these can cause coloration of the water bodies affecting natural photosynthesis of algae and marine plants disrupting the aquatic food chain. The remedial methods for dye contamination are of many types. These could be of chemical means which produce OH free radicals for the oxidation of dyes [2] or reagents like ozone which too cause oxidation. Further, these could be of biological means where usage of a microbial fuel cell [3] can be seen. However, the most pursued method is of adsorption where a substrate adsorbs the dye from the water via interactions like electrostatic attractions. Many substrates are used for the same as bentonite clays [4], activated carbon [5], etc. But the choice of substrate should be based on sustainability and green principles for environment benign removal. Thus, the use of natural adsorbents like chitin and cellulose is highly researched nowadays. Cellulose is the most abundant natural polymer and can be obtained from trees, weeds [6–8], and even agricultural wastes [9]. Figure 2.1 shows the different sources for nanocellulose.

Cellulose can be obtained from a multitude of plant sources like leaves, fruit hulls [11], wood, etc. apart from agricultural wastes and weeds. Nowadays, it is also extracted from microbes like algae and bacteria [12,13] called as bacterial cellulose and sea animals like tunicates [14]. The only hurdle is that the amount of cellulose that can be obtained from each different source varies. But plants can be considered as an inexhaustible resource and as such cellulose as polymer can be considered inexhaustible as opposed to other polymers that are obtained from fossil fuel sources. Considering agricultural wastes as a source means that the cost gets lowered and that sustainability is being put

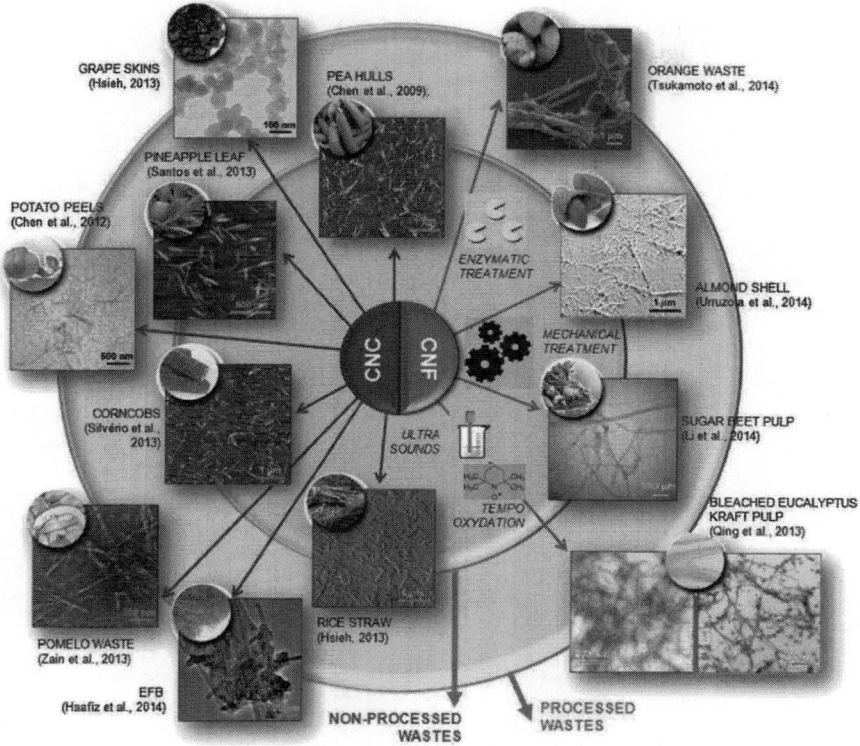

Figure 2.1   Different sources of nanocellulose (microscopy images of CNCs and CNFs obtained from different sources). Reproduced with the permission from Elsevier Ltd [10]

into practice. Based on this principle many research groups have worked on using agricultural wastes like soy hulls and rice husk as sources and extracting cellulose from them.

2.2   Extraction of cellulose nanomaterials

Cellulose in plants is present along with lignin and hemicellulose where lignin and hemicellulose act as a matrix in which cellulose is embedded as shown in Fig. 2.2 [15]. These must be removed while extracting cellulose and for the same, many methods have been tried.

Earlier, mechanical methods were applied which were highly energy consuming. However, many other researchers tried pretreatments to remove lignin and hemicellulose early on and then go for mechanical treatment. The pretreatments include alkali treatment [17], enzymatic treatment [18], etc. However, to obtain cellulose nano fibril (CNF), a mechanical treatment is

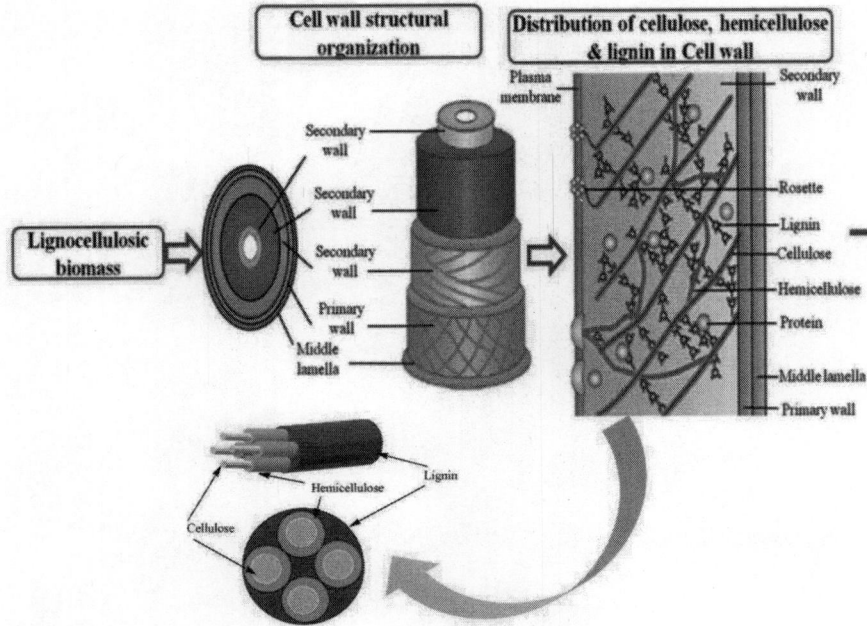

Figure 2.2 Diagrammatic illustration of the framework of lignocellulose, cellulose, hemicellulose and lignin. Reproduced with the permission from Elsevier Ltd [16]

irreplaceable which can be homogenization, ultrasonication, steam explosion electrospinning. Chen et al. [19] reported a method to fibrillate the raw dried cotton fibers into individual cellulose nanofibers (CNFs) by chemical purification and pretreatment by a high-speed blender combined with high pressure homogenization. Abe et al. [20] obtained CNFs with a uniform width of 15 nm from wood by the grinding treatment in an undried state. Cryo crushing combined with a high-pressure fibrillation process was used by Wang and Sain [21] for the isolation of CNFs with diameters in the range of 50–100 nm from soybean stock. Deepa et al. [22] reported the extraction of CNFs from the banana plant by the steam explosion process in an autoclave. Tibolla et al. [23] isolated CNFs from banana peel bran using chemical treatment and enzymatic treatment (alkaline treatment and hydrolysis with xylanase). Nanofibers produced from enzymatic treatment had an average diameter of 7.6 nm and length in the range of 2889.7 nm [23].

Cellulose nanocrystals (CNCs) are typically prepared by acid hydrolysis of the CNFs as shown in Fig. 2.3. The amorphous portion in the cellulose microfibers could be hydrolyzed via strong acid treatment and the resulting rod like crystalline structures are known as CNCs or whiskers. Numerous types

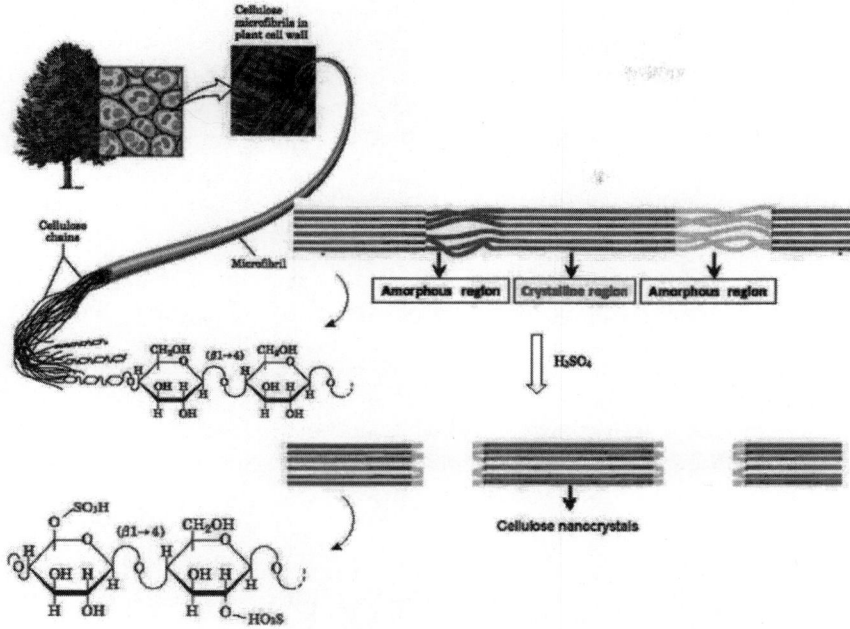

Figure 2.3 Schematic diagram illustrating the hierarchy of cellulose biosynthesis and the sulfuric acid hydrolysis of cellulose to produce sulphated CNCs. Reproduced with the permission from Elsevier Ltd [24]

of acids have been used for the extraction of CNCs which includes sulphuric acid, hydrochloric acid, oxalic acid, hydrobromic acid (HBr) and nitric acid. Among the acid hydrolysis treatments, sulphuric acid treatment of cellulosic material has been widely used to extract CNCs. Danial et al. [25] extracted the CNCs from waste paper by acid hydrolysis using 60% (V/V) sulphuric acid solution at 45 °C with constant stirring. They obtained the CNCs with length ranged in 100–300 nm [25]. Acid hydrolysis can cause these regions to be separated out and the crystalline regions are called as CNCs.

2.3   Comparison between CNFs and CNCs

Cellulose nanocrystals and CNF are very similar. The major difference is that CNCs are more crystalline and are shorter than CNFs. The higher length of CNF results in percolation and entanglements which can cause enhanced reinforcement property. Nanocellulose can be classified into CNFs, CNCs and bacterial nanocellulose. Among the above mentioned nanocellulose structures, CNFs and CNCs gained considerable attention due to its excellent mechanical properties and ease of production. The major difference between the CNCs and CNFs lies in their amorphous portions and dimensions. The CNCs have nano dimensions in both length and diameter wise, whereas CNFs have length in micro dimension and diameter is in the nano dimensions as shown in Fig 2.3. Cellulose nano fibril can be manufactured by various chemo-mechanical treatments, whereas CNCs by strong acid hydrolysis using sulphuric acid and it destroys all the amorphous portion (dis-ordered region) and lead to the nanocrystal structure. Both CNCs and CNFs can be easily extracted from the plant cell walls.

2.4   Various processable forms of nanocellulose

Cellulose nanofibers and CNCs both can be used as a reinforcing agent, but they can be applied as a polymeric matrix also. Nanocellulose could be transformed into various processable forms like nanocellulose hydrogel, cellulose nano paper, nanocellulose aerogel. As shown in Fig. 2.4, the

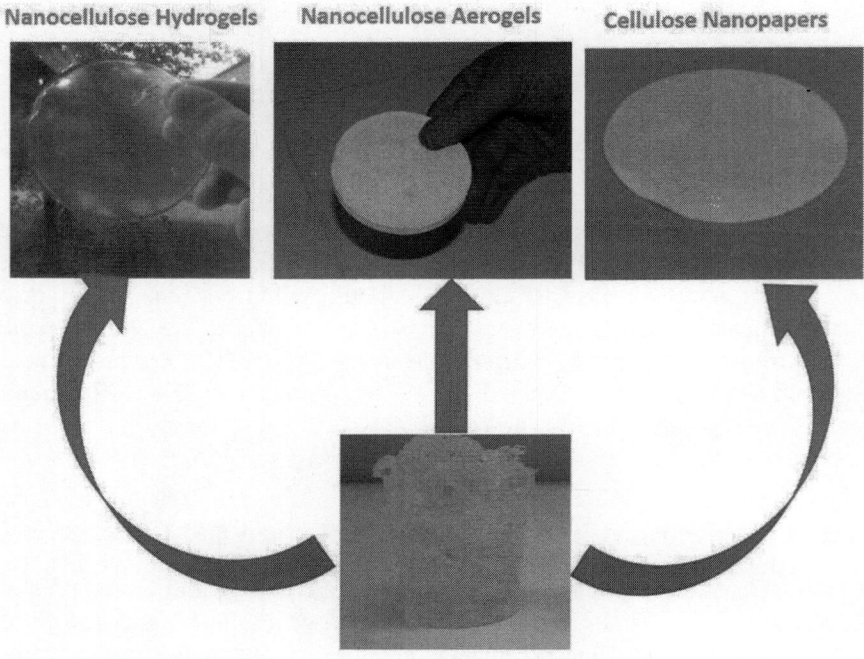

Figure 2.4   Images of nanocellulose dispersion (a), nanocellulose hydrogel (b), and atypical porous cellulose nanopaper (c). Reproduced and modified with the permission from American Chemical Society [26]

nanocellulose suspension can be processed into various forms depending upon the application.

Normally nanocellulose hydrogels could be produced from the nanocellulose dispersion via vacuum filtration process. Initially the cellulose nanosuspension could be diluted to 0.5 wt% and then filtrated on the top of the filter membrane. By this way one can easily produce nanocellulose hydrogel as shown in Fig. 2.4 from the nanocellulose dispersion. Nanocellulose hydrogel have many potential applications especially in bio-medical field. It can be used as scaffolds for wound healing, tissue engineering, and carrier for drugs. Nanocellulose aerogel is another processable form of the nanocellulose. Nanocellulose aerogel can be fabricated via simple freeze-drying process of the 3 wt% of nanocellulose dispersion. Due to its light weight, high bulk volume, mechanical strength, high porous nature, it has gained in the considerable interest among the scientific community. Recently a lot of researches have been reported on the excellent oil absorption capacities of the nanocellulose aerogel. Since nanocellulose aerogel inherent hydrophilic character limits its application in various fields, scientists around the world have been started to induce the hydrophobicity in the nanocellulose aerogels for various applications. Moreover, nanocellulose can be used for water purification, thermal insulation, air filtration, electromagnetic shielding, etc., on the other hand, cellulose nanopapers are the important processable form of the nanocellulose as shown in Fig. 2.4. Cellulose nanopapers are a class of promising functional constructs that can be exploited for varying technological applications ranging from flexible optoelectronic devices [27–29] to nonwoven porous membranes for water purification [30,31] and as fire-retarding gas barrier films [32,33]. Such applications are made possible due to superior mechanical properties, high aspect ratio, thermal stability, and chemical resistance of nanocellulose fibers, all of which can be sensibly fine-tuned as per the application.

2.5   Problem with toxic textile dyes

Dyes are substances used to impart coloration to different objects like textiles. A significant property of dyes is that even a small amount can cause coloration. Among them, synthetic dyes are even more problematic. They have lesser biodegradability and thus usually end up being dumped into the hydrosphere. The consequence of dyes in water is that usually these compounds are carcinogenic and via the method of bio magnification, these can be found in higher animals including humans. Further, these can also have specific problems mutagenic and teratogenic in several fish species [34]. In humans these are responsible for kidney dysfunction, damage to central nervous system, and reproductive system [35]. Apart from these there is another implication viz coloring of water bodies thereby decreasing the sunlight penetration. The coloration disrupts the photosynthetic activity of aquatic plants and microbes, which usually form the basis of an aquatic food chain. So, the aquatic ecosystem maybe disturbed. Aquatic fauna is also affected. This is because the presence of any pollutants usually ends up increasing the bio chemical oxidation demand of water. This means the amount of oxygen present dissolved in water, available to the fauna is decreased considerably. This also further escalates the disruption of aquatic ecosystem.

2.6   Relevance of nanocellulose in water purification

Cellulose has a surface hydroxyl group which can be substituted for different functionalities. Considering the problems of dyes, cellulose can be used as an adsorbent for them owing to its negative charge. Dyes can be mainly classified as cationic or anionic dyes depending upon the chemical groups in them. For cationic dyes, cellulose will act as a good adsorbent. But for anionic dyes, the surface OH must be functionalized to hold a positive charge. Thus, it can be used for removal of both positively charged particles. This is in comparison to other polymers or substrates which are derived from exhaustible sources like fossil fuel. Thus, cellulose has multifunctionality along with sustainability and cheapness.

Nanocellulose can be tailored to exhibit novel and significantly improved physical and chemical properties. The unique features of nanocellulose include small diameter, high surface-to-volume ratio, abundant hydroxyl groups for easy functionalization, good mechanical properties, and good chemical resistance. These excellent properties make nanocellulose as a huge potential candidate for waste water treatment. In an aqueous environment, the high surface area and high aspect ratio of nanocellulose are beneficial for the formation of an ultrafine three-dimensional network structures which can be explored for removal and absorption of various pollutants in the water. Since cellulose is hydrophilic in nature, cellulose has been used as an antifouling hydrophilic coating to increase the flux of the membranes. Cellulose tends to exhibit a high adsorption capacity for pollutant after suitable chemical modification on its surface with the aim of incorporating molecules that contain basic groups, particularly those that are rich in nitrogen, sulfur, and oxygen [36]. Cellulose nanofibers have been intensively studied due to their natural abundance, ease of functionalization, and structural diversity [37,38]. Nanotechnologies have been touted as having great potential for reducing costs and improving efficiency in pollution prevention, treatment, and clean up. Nanocellulose has generated interest in this area as an active sorbent material for contaminants and as stabilizer for other active particles. The easily functionalizable surface of nanocellulose allows for the incorporation of different chemical moieties that may enhance the binding and adsorption efficiency against the pollutants in water such as dyes, toxic heavy metals, etc. Cellulose nanomaterials (CNs) are a promising alternative adsorbent due to its high surface area-to-volume ratio, low cost, high natural abundance, and inherent environmental inertness. Moreover, CN's easily functionalizable surface allows for the incorporation of chemical moieties that may increase the binding efficiency of pollutants to the CN. Carboxylation of CNs is by far the most studied method for increasing their sorptive capacity. The sorption of organic contaminants has also been demonstrated with modified CN matrices. The inherent hydrophilicity of CNs can be reduced to improve the affinity of the material for hydrophobic compounds. The manipulation of the surface chemistry of CNs can be achieved by inclusion of both organic and inorganic functionalities. The rapid progress made in the nanosciences, especially with nanocellulose; offer a potential new type of material for use in water treatment in the form of nanoparticles. Nanocellulose is efficient absorbent materials due to its high surface area-to-volume ratio, low cost, high natural abundance, and inherent environmental inertness. The manipulation of the surface chemistry of nanocellulose can be achieved by inclusion of both organic and inorganic functionalities; thereby it acts as a water repellent such that it could remove oils that were spread on the surface of the water.

Nanocellulose based microfiltration membranes for water treatment has gained considerable interest nowadays. The dimensions of nanocellulose and the strength of the material can be exploited in the fabrication of membranes for water treatment. Membranes can greatly benefit from the high strength of nanocellulose. At low loadings of a few wt% nanocellulose can increase the tensile strength of polymer membrane up to 50% [39,40]. At low nanocellulose loading (<4 wt%), membranes show increased porosity, larger pore size, and greater surface hydrophilicity which lead to greater water permeability. The addition of nanocellulose to polymer membrane has generally been shown to increase membrane hydrophilicity, thereby reducing the antifouling problems of the membrane. All of these properties include changes in the membrane surface hydrophilicity, greater permeability, greater selectivity, and greater resistance to biofouling that make nanocellulose as a promising candidate for waste water treatment applications.

2.7   Nanocellulose as efficient adsorbents against toxic textile dyes

The use of synthetic chemical dyes has increased considerably in the paper, textile, food, and pharmaceutical industries over the past few years, resulting in the release and accumulation of dye-containing industrial effluents into the aquatic systems. It is estimated that there are more than 100,000 types of commercially available dyes and an annual worldwide production of 700,000–1,000,000 tons have been reported in Ref. [41]. Since most of these dyes are toxic in nature, their presence in water is one of the major environmental concerns because they are usually very recalcitrant to microbial degradation. Most of the dyes have complex aromatic structure resistant to light, biological activity, and ozone and not readily removed by typical waste treatment processes. Adsorption is the one of the efficient methods to remove textile dyes from the water. In recent years, investigations have been undertaken to evaluate inexpensive alternate materials of biological origin as potential adsorbent for dyes. Among this nanocellulose from bio-mass has gained considerable attention due to its readily functionalizable surface, natural abundance, and high strength. Some literatures have been reported on the dye removal from waste water via adsorption using nanocellulose.

Ma et al. [42] reported the removal of crystal violet dyes using nanofibrous microfiltration membrane based on cellulose nanowhiskers via adsorption. They have done TEMPO oxidation on the cellulose nanowhiskers surface. The cellulose nanowhiskers based nanofibrous microfiltration membrane was prepared by impregnating cellulose nanowhiskers into the poly(acrylonitrile) (PAN) electrospun scaffold, resulting in a cross-linked nanofibrous mesh with very high surface-to-volume ratio. They found that the maximum adsorption capacity of the cellulose nanowhisker-based microfiltration membrane for crystal violet (CV) dye was 16 times higher than that of GS0-22 (Commercially available membrane) shown in Fig. 2.5. This was due to the fact that the strong electrostatic attraction between protonated CV molecules and highly negative carboxylate group on the cellulose nanowhiskers surface [42].

Beyki et al. [43] removed Congo red dye by using an efficient magnetic Fe_3O_4 cellulose core. They synthesized magnetic Fe_3O_4 cellulose core by one step method. This magnetic Fe_3O_4 cellulose has further reacted with epichlorohydrin and 1-methylimidazole to generate magnetic polymeric ionic liquid. Their results revealed that different factors like contact time, adsorbent dyes, and ionic strength have significant effect on Congo red dye adsorption. Regeneration study showed that this demonstrated nanohybrid had good reusability as a magnetic sorbent for environmental remediation purpose [43]. Jin et al. [44] investigated the amino-functionalized nanocrystalline cellulose as an adsorbent for anionic dyes. They showed that the amine-functionalized nanocrystalline cellulose had the high adsorption removal of Congo red dyes (about 100%). This was mainly due to the strong electrostatic and chemical interaction between the amine groups present on Congo red dye and amine and hydroxyl groups on amino-functionalized nanocrystalline cellulose [44]. Chong et al. [45] also removed Congo red dyes from the aqueous solution using $CaCO_3$-decorated cellulose aerogels. The $CaCO_3$-decorated cellulose aerogel was prepared via insitu precipitation of $CaCO_3$ into as-prepared cellulose aerogel, which was prepared via freeze-drying. Adsorption studies illustrated that the

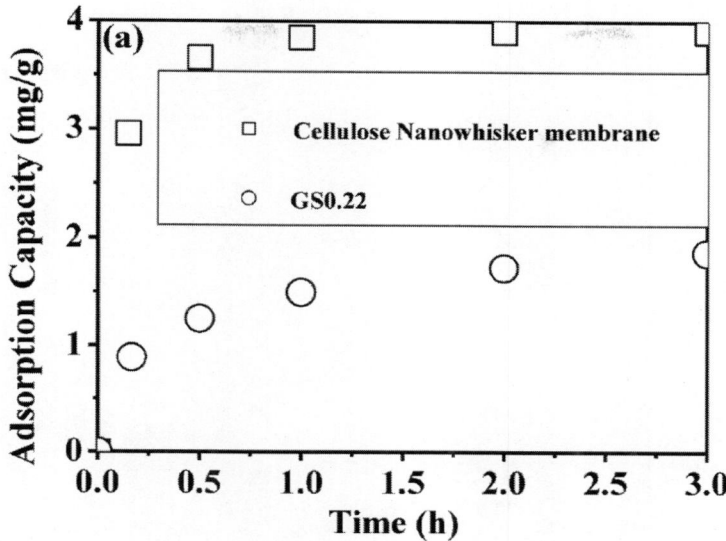

Figure 2.5 Adsorption capacity towards crystal violet dyes (CV) of cellulose nanowhisker-based nanofibrous MF membrane and GS0.22 against time [42]. Reproduced with permission from American Chemical Society

Congo red uptake by the aerogel was dependent on the dye concentration and temperature. The maximum adsorption capacity of the $CaCO_3$-cellulose aerogel towards Congo red dye was approximately 75.81 mg/g [45]. Jin et al. [46] demonstrated a novel nanocomposite microgels based on nanocellulose and amphoteric polyvinyl amine (PVAm) via two-step method for the removal of anionic dyes from aqueous solutions. The fabricated microgel was found to be effective in anionic dye removal at acidic conditions due to the protonation of amino groups. Absorption studies revealed that the fabricated microgel had high adsorption towards Congo red dye which was 1469 mg/g [46].

Many industries especially textile industry, paper industry, and plastics industry consume a large amount of dyes in order to color their products. Most of these dyes are water soluble and they can contaminate water easily. This dye contamination can lead to serious environmental and health issues. However, the dye polluted water is very difficult to treat because of recalcitrant nature of dyes and their resistance towards light, heat, and oxidizing agents. These organic dyes possess a complex aromatic structure and can exhibit cationic, anionic, and non-ionic nature.

Batmaz et al. [47] explored the possibility of carboxylated CNCs as a sorbent for cationic dyes. In this study, they incorporated carboxylic group by TEMPO oxidation of cellulose. They could observe that the absorbance performance of the carboxylated CNC is comparable to activated carbons. The chart in Fig. 2.6

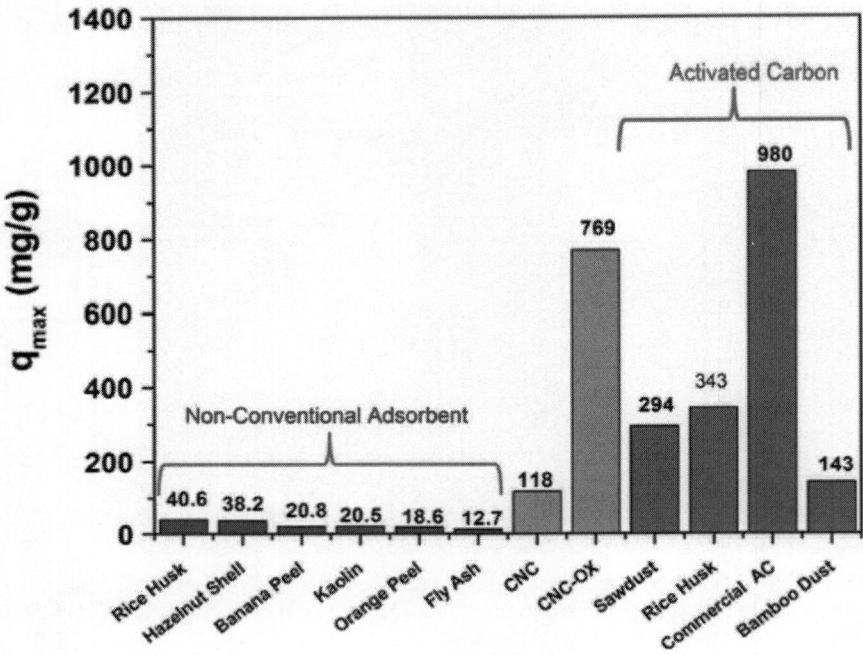

Figure 2.6 Comparison of of various adsorbents [47]. Reproduced with the permission from Springer

clearly illustrates the efficiency of both CNC and carboxylated CNC over other non-conventional adsorbents.

In a recent report, Yu et al. introduced a new approach for single step extraction of CNCs and their efficiency in cationic dye removal. They could extract carboxylated CNC with typical rod like structure using novel $C_6H_8O_7$/ HCl hydrolysis of microcrystalline cellulose. They also demonstrated the efficiency of this extracted CNC (92.3% at 20 mg/mL adsorbent in initial dye concentration of 200 mg/L) was better than that of carboxylated CNCs prepared by TEMPO oxidation as shown in Fig. 2.7 [48].

Manna et al. [49] reported a novel method to functionalize lignocellulosic fibers using neem oil–phenolic resin emulsion. The treated lignocellulosic fibers exhibited rapid adsorption of methylene blue over a wide range of pH conditions.

In this study, Manna et al. showed that the rapid methylene blue adsorption was due to the coupled effect of electrostatic interactions, van der Waals forces, and hydrogen bonding interactions between cationic methylene blue and treated fiber as shown in Fig. 2.8 [49]. Pei et al. [50] reported surface quaternization of CNFs for making nanopapers with high anionic dye removal ability. Divalent anionic dyes, Congo red and acid green 25 were utilized for

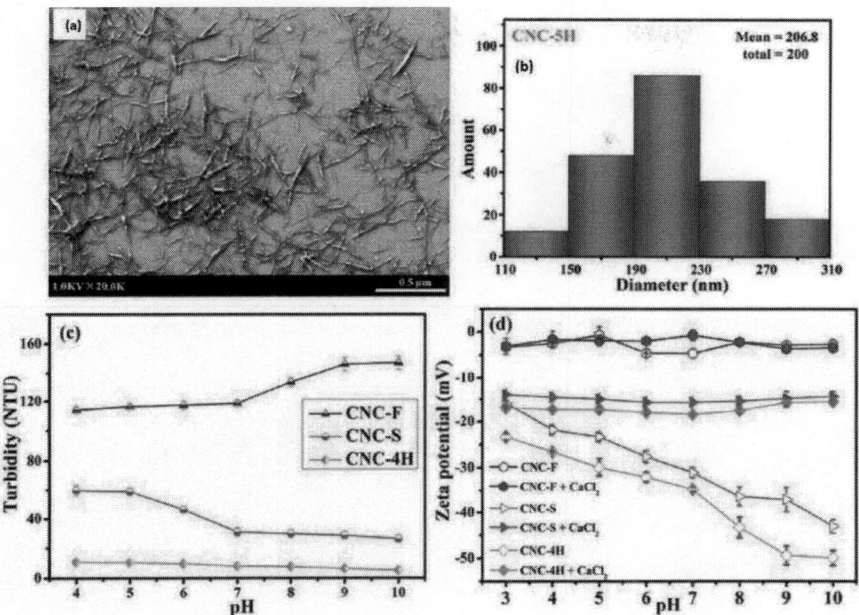

Figure 2.7   (a) FESEM image of extracted CNC (b) size distribution of extracted CNC (c) effect of three kinds of CNC dosage on dye removal (%) and (d) effect of pH on dye removal for three kinds of CNCs (Inset photos of dye before and after adsorption with CNCs at pH=7) [48]. Reproduced with the permission from Springer

dye adsorption studies. They could observe a linear relationship between dye adsorption and triethylammonium content from 0.59 to 1.32 mmol/g. Further increase in triethylammonium content could not generate a significant increase in dye adsorption as shown in Fig. 2.9 [50].

Recently Gopakumar et al. [51] removed the toxic crystal violet dyes from water by using CNF based polyvinylidene fluoride (PVDF) membrane via Meldrum's acid modification of CNFs. They introduced a facile method to produce a unique green adsorbent material from CNFs via a nonsolvent assisted procedure using Meldrum's acid as an esterification agent to enhance the adsorption toward positively charged crystal violet dyes. They showed that, with 10 mg/L of crystal violet (CV) aqueous solution, CV adsorption of PVDF electrospun membrane, and unmodified CNF-based PVDF membrane was around 1.368 and 2.948 mg/g of the membrane respectively, whereas it was 3.984 mg/g of the membrane by Meldrum's acid CNF-based PVDF membrane as shown in Fig. 2.10. They found that the enhanced absorption capacity of Meldrum's acid modified membrane is due to the electrostatic interaction between the protonated crystal violet and Meldrum's acid modified CNFs as shown in Fig. 2.11 [51].

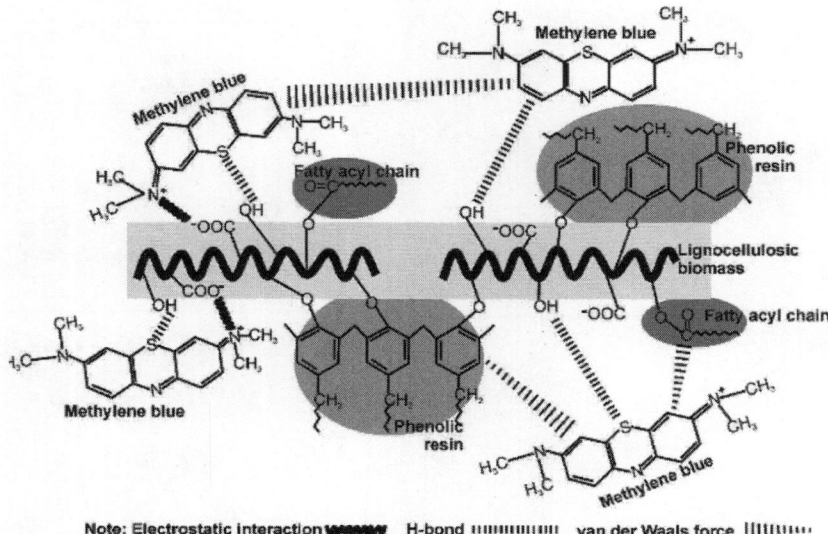

Figure 2.8 Interaction of methylene blue with the treated fiber. Reproduced with the permission from Elsevier Ltd [49]

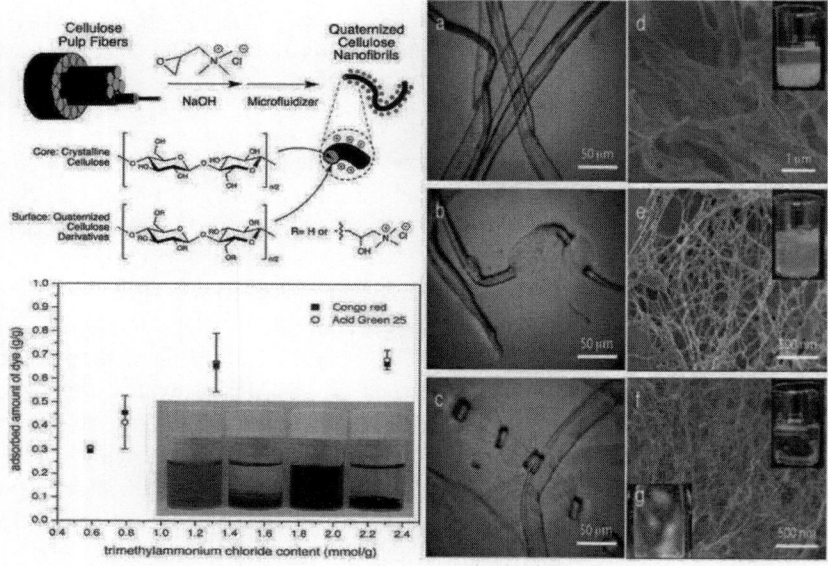

Figure 2.9 Schematic representation of quaternization of cellulose (Upper right). (a–c) These are phase contrast micrographs of quarternized fibers having different trimethyl-ammonium chloride content. (d–f) These are FESEM images of quarternized cellulose fibers [50]. Reproduced with the permission from Royal Society of Chemistry

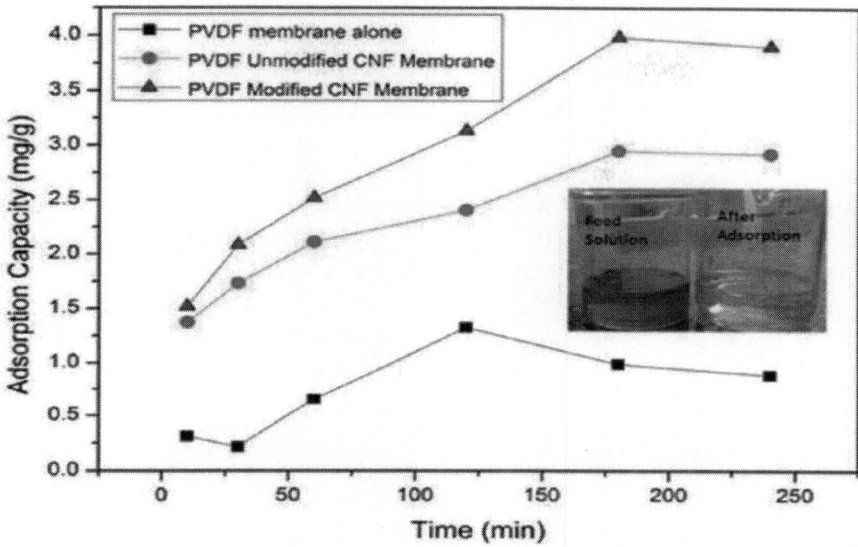

Figure 2.10 Adsorption capacity of PVDF membrane alone, PVDF/unmodified CNF membrane, and PVDF/modified CNF membranes against time. Reproduced with permission from American Chemical Society [51]

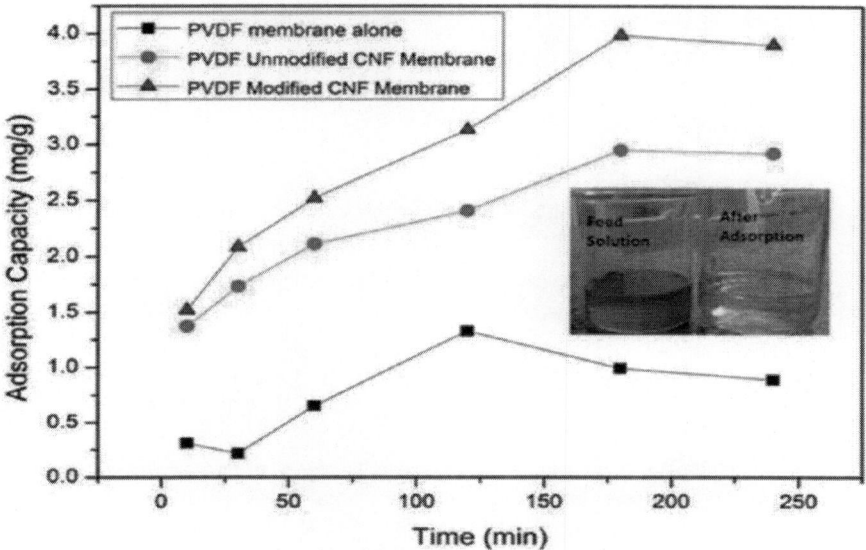

Figure 2.11 Schematic representation of crystal violet removal by Meldrum's acid-modified CNF-based PVDF nanofibrous MF membrane and its mechanism. Reproduced with permission from American Chemical Society [51]

2.8   Conclusions

Advances in industrial and agriculture sector have led to generation of huge quantity of waste water containing toxic pollutants. Among these pollutants, textile dyes are one of the larger group of pollutants which are released from textile industrial waste water. These toxic textile dyes have adverse effects on the human and aquatic life. So, there is every need to formulate the new techniques for the removal of these toxic textile dyes from the water. In this context, several researches have been attempted to remove the toxic textile dyes by using the different bio-absorbents. Due to the increase of environmental pollution, scientists globally have opted for biodegradable nanomaterials as a new generation of adsorbents. Cellulose is the most abundant source in the Earth, cellulose has been intensively studied by the scientists all over the world. Nanocellulose inherent fibrous nature and remarkable properties, suggest as a promising candidate for water filtration membranes. Cellulose nanomaterials are a promising adsorbent against textile dyes due to its high surface area-to-volume ratio, low cost, and natural abundance. Cellulose nanomaterials easily functionalizable surface allows different surface modifications that may increase the binding efficiency of pollutants to the CNs surface. The presence of high surface density of hydroxyl groups, a broad possibility of surface modification based on the chemistry of hydroxyl groups is possible to implement, opening the way toward selective adsorption towards various kinds of anionic and cationic textile dyes.

References

1. Carneiro, P. A.; Nogueira, R. F. P.; Zanoni, M. V. B. Homogeneous photodegradation of C.I. Reactive Blue 4 using a photo-Fenton process under artificial and solar irradiation. *Dye Pigment.* **2006**, doi:10.1016/j.dyepig.2006.01.022.

2. Basturk, E.; Karatas, M. Decolorization of antraquinone dye Reactive Blue 181 solution by UV/H$_2$O$_2$ process. *J. Photochem. Photobiol. A Chem.* **2015**, doi:10.1016/j.jphotochem.2014.11.003.

3. Fang, Z.; Song, H. L.; Cang, N.; Li, X. N. Electricity production from Azo dye wastewater using a microbial fuel cell coupled constructed wetland operating under different operating conditions. *Biosens. Bioelectron.* **2015**, doi:10.1016/j.bios.2014.12.047.

4. Anirudhan, T. S.; Ramachandran, M. Adsorptive removal of basic dyes from aqueous solutions by surfactant modified bentonite clay (organoclay): Kinetic and competitive adsorption isotherm. *Process Saf. Environ. Prot.* **2015**, doi:10.1016/j.psep.2015.03.003.

5. Maneerung, T.; Liew, J.; Dai, Y.; Kawi, S.; Chong, C.; Wang, C. H. Activated carbon derived from carbon residue from biomass gasification and its application for dye adsorption: Kinetics, isotherms and thermodynamic studies. *Bioresour. Technol.* **2016**, doi:10.1016/j.biortech.2015.10.047.

6. Manna, S.; Gopakumar, D. A.; Roy, D.; Saha, P.; Thomas, S. Nanobiomaterials for removal of fluoride and chlorophenols from water. In *New Polymer Nanocomposites for Environmental Remediation*; **2018** ISBN 9780128110348.

7. Gopakumar, D. A.; Thomas, S.; Grohens, Y. Nanocelluloses as innovative polymers for membrane applications. In *Multifunctional Polymeric Nanocomposites Based on Cellulosic Reinforcements*; **2016**; pp. 253–275 ISBN 9780323442480.

8. Gopakumar, D. A.; Pai, A. R.; Pottathara, Y. B.; Pasquini, D.; Carlos De Morais, L.; Luke, M.; Kalarikkal, N.; Grohens, Y.; Thomas, S. Cellulose nanofiber-based polyaniline flexible papers as sustainable microwave absorbers in the X-Band. *ACS Appl. Mater. Interfaces* **2018**, doi:10.1021/acsami.8b04549.

9. Bhatnagar, A.; Sillanpää, M.; Witek-Krowiak, A. Agricultural waste peels as versatile biomass for water purification – A review. *Chem. Eng. J.* **2015**, doi:10.1016/j.cej.2015.01.135.

10. García, A.; Gandini, A.; Labidi, J.; Belgacem, N.; Bras, J. Industrial and crop wastes: A new source for nanocellulose biorefinery. *Ind. Crops Prod.* **2016**, doi:10.1016/j.indcrop.2016.06.004.

11. Rosa, M. F.; Medeiros, E. S.; Malmonge, J. A.; Gregorski, K. S.; Wood, D. F.; Mattoso, L. H. C.; Glenn, G.; Orts, W. J.; Imam, S. H. Cellulose nanowhiskers from coconut husk fibers: Effect of preparation conditions on their thermal and morphological behavior. *Carbohydr. Polym.* **2010**, doi:10.1016/j.carbpol.2010.01.059.

12. Lin, W. C.; Lien, C. C.; Yeh, H. J.; Yu, C. M.; Hsu, S. H. Bacterial cellulose and bacterial cellulose-chitosan membranes for wound dressing applications. *Carbohydr. Polym.* **2013**, doi:10.1016/j.carbpol.2013.01.076.

13. Chen, Y. W.; Lee, H. V.; Juan, J. C.; Phang, S. M. Production of new cellulose nanomaterial from red algae marine biomass Gelidium elegans. *Carbohydr. Polym.* **2016**, doi:10.1016/j.carbpol.2016.06.083.

14. Zhao, Y.; Li, J. Excellent chemical and material cellulose from tunicates: Diversity in cellulose production yield and chemical and morphological structures from different tunicate species. *Cellulose* **2014**, doi:10.1007/s10570-014-0348-6.

15. Rong, M. Z.; Zhang, M. Q.; Liu, Y.; Yang, G. C.; Zeng, H. M. The effect of fiber treatment on the mechanical properties of unidirectional sisal-reinforced epoxy composites. *Compos. Sci. Technol.* **2001**, doi:10.1016/S0266-3538(01)00046-X.

16. Anwar, Z.; Gulfraz, M.; Irshad, M. Agro-industrial lignocellulosic biomass a key to unlock the future bio-energy: A brief review. *J. Radiat. Res. Appl. Sci.* **2014**, doi:10.1016/j.jrras.2014.02.003.

17. Zuluaga, R.; Putaux, J. L.; Cruz, J.; Vélez, J.; Mondragon, I.; Gañán, P. Cellulose microfibrils from banana rachis: Effect of alkaline treatments on structural and morphological features. *Carbohydr. Polym.* **2009**, doi:10.1016/j.carbpol.2008.09.024.

18. Nie, S.; Zhang, K.; Lin, X.; Zhang, C.; Yan, D.; Liang, H.; Wang, S. Enzymatic pretreatment for the improvement of dispersion and film properties of cellulose nanofibrils. *Carbohydr. Polym.* **2018**, doi:10.1016/j.carbpol.2017.11.020.

19. Chen, W.; Abe, K.; Uetani, K.; Yu, H.; Liu, Y.; Yano, H. Individual cotton cellulose nanofibers: Pretreatment and fibrillation technique. *Cellulose* **2014**, *21*, 1517–1528, doi:10.1007/s10570-014-0172-z.

20. Abe, K.; Iwamoto, S.; Yano, H. Obtaining cellulose nanofibers with a uniform width of 15 nm from wood. *Biomacromolecules* **2007**, *8*, 3276–3278, doi:10.1021/bm700624p.

21. Wang, B.; Sain, M. Isolation of nanofibers from soybean source and their reinforcing capability on synthetic polymers. *Compos. Sci. Technol.* **2007**, *67*, 2521–2527, doi:10.1016/j.compscitech.2006.12.015.

22. Deepa, B.; Abraham, E.; Cherian, B. M.; Bismarck, A.; Blaker, J. J.; Pothan, L. A.; Leao, A. L.; de Souza, S. F.; Kottaisamy, M. Structure, morphology and thermal characteristics of banana nano fibers obtained by steam explosion. *Bioresour. Technol.* **2011**, *102*, 1988–1997, doi:10.1016/j.biortech.2010.09.030.

23. Tibolla, H.; Pelissari, F. M.; Rodrigues, M. I.; Menegalli, F. C. Cellulose nanofibers produced from banana peel by enzymatic treatment: Study of process conditions. *Ind. Crops Prod.* **2017**, doi:10.1016/j.indcrop.2016.11.035.

24. Koshani, R.; Madadlou, A. A viewpoint on the gastrointestinal fate of cellulose nanocrystals. *Trends Food Sci. Technol.* **2018**, doi:10.1016/j.tifs.2017.10.023.

25. Danial, W. H.; Abdul Majid, Z.; Mohd Muhid, M. N.; Triwahyono, S.; Bakar, M. B.; Ramli, Z. The reuse of wastepaper for the extraction of cellulose nanocrystals. *Carbohydr. Polym.* **2015**, doi:10.1016/j.carbpol.2014.10.072.

26. Sehaqui, H.; Zhou, Q.; Ikkala, O.; Berglund, L. A. Strong and tough cellulose nanopaper with high specific surface area and porosity. *Biomacromolecules* **2011**, doi:10.1021/bm2008907.

27. Lizundia, E.; Delgado-Aguilar, M.; Mutjé, P.; Fernández, E.; Robles-Hernandez, B.; de la Fuente, M. R.; Vilas, J. L.; León, L. M. Cu-coated cellulose nanopaper for green and low-cost electronics. *Cellulose* **2016**, doi:10.1007/s10570-016-0920-3.

28. Yagyu, H.; Saito, T.; Isogai, A.; Koga, H.; Nogi, M. Chemical modification of cellulose nanofibers for the production of highly thermal resistant and optically transparent nanopaper for paper devices. *ACS Appl. Mater. Interfaces* **2015**, *7*, 22012–22017, doi:10.1021/acsami.5b06915.

29. Hsieh, M. C.; Kim, C.; Nogi, M.; Suganuma, K. Electrically conductive lines on cellulose nanopaper for flexible electrical devices. *Nanoscale* **2013**, doi:10.1039/c3nr01951a.

30. Mautner, A.; Lee, K. Y.; Tammelin, T.; Mathew, A. P.; Nedoma, A. J.; Li, K.; Bismarck, A. Cellulose nanopapers as tight aqueous ultra-filtration membranes. *React. Funct. Polym.* **2015**, *86*, 209–214, doi:10.1016/j.reactfunctpolym.2014.09.014.

31. Sehaqui, H.; Morimune, S.; Nishino, T.; Berglund, L. A. Stretchable and strong cellulose nanopaper structures based on polymer-coated nanofiber networks: An alternative to nonwoven porous membranes from electrospinning. *Biomacromolecules* **2012**, *13*, 3661–3667, doi:10.1021/bm301105s.

32. Liu, A.; Walther, A.; Ikkala, O.; Belova, L.; Berglund, L. A. Clay nanopaper with tough cellulose nanofiber matrix for fire retardancy and gas barrier functions. *Biomacromolecules* **2011**, *12*, 633–641, doi:10.1021/bm101296z.

33. Nair, S. S.; Zhu, J.; Deng, Y.; Ragauskas, A. J. High performance green barriers based on nanocellulose. *Sustain. Chem. Process.* **2014**, *2*, 1–7, doi:10.1186/s40508-014-0023-0.

34. Yagub, M. T.; Sen, T. K.; Afroze, S.; Ang, H. M. Dye and its removal from aqueous solution by adsorption: A review. *Adv. Colloid Interface Sci.* **2014**, doi:10.1016/j. cis.2014.04.002.

35. Kadirvelu, K.; Kavipriya, M.; Karthika, C.; Radhika, M.; Vennilamani, N.; Pattabhi, S. Utilization of various agricultural wastes for activated carbon preparation and application for the removal of dyes and metal ions from aqueous solutions. *Bioresour. Technol.* **2003**, doi:10.1016/S0960-8524(02)00201-8.

36. Musyoka, S. M.; Ngila, J. C.; Moodley, B.; Petrik, L.; Kindness, A. Synthesis, characterization, and adsorption kinetic studies of ethylenediamine modified cellulose for removal of Cd and pb. *Anal. Lett.* **2011**, doi:10.1080/00032719.2010.539736.

37. Fernandes, S. C. M.; Freire, C. S. R.; Silvestre, A. J. D.; Pascoal Neto, C.; Gandini, A. Novel materials based on chitosan and cellulose. *Polym. Int.* **2011**, doi:10.1002/pi.3024.

38. Peterson, J. J.; Willgert, M.; Hansson, S.; Malmström, E.; Carter, K. R. Surface-Grafted conjugated polymers for hybrid cellulose materials. *J. Polym. Sci. Part A Polym. Chem.* **2011**, doi:10.1002/pola.24733.

39. Lalia, B. S.; Guillen, E.; Arafat, H. A.; Hashaikeh, R. Nanocrystalline cellulose reinforced PVDF-HFP membranes for membrane distillation application. *Desalination* **2014**, doi:10.1016/j.desal.2013.10.030.

40. Qu, P.; Tang, H.; Gao, Y.; Zhang, L. P.; Wang, S. Polyethersulfone composite membrane blended With cellulose fibrils. *BioResources* **2010**, doi:10.1016/j.sbspro.2014.01.1130.

41. Pang, Y. L.; Abdullah, A. Z. Current status of textile industry wastewater management and research progress in malaysia: A review. *Clean – Soil, Air, Water* **2013**, doi:10.1002/clen.201000318.

42. Ma, H.; Burger, C.; Hsiao, B. S.; Chu, B. Nanofibrous microfiltration membrane based on cellulose nanowhiskers. *Biomacromolecules* **2012**, *13*, 180–186, doi:10.1021/bm201421g.

43. Beyki, M. H.; Bayat, M.; Shemirani, F. Fabrication of core-shell structured magnetic nanocellulose base polymeric ionic liquid for effective biosorption of Congo red dye. *Bioresour. Technol.* **2016**, doi:10.1016/j.biortech.2016.06.069.

44. Jin, L.; Li, W.; Xu, Q.; Sun, Q. Amino-functionalized nanocrystalline cellulose as an adsorbent for anionic dyes. *Cellulose* **2015**, doi:10.1007/s10570-015-0649-4.

45. Chong, K. Y.; Chia, C. H.; Zakaria, S.; Sajab, M. S.; Chook, S. W.; Khiew, P. S. CaCO3-decorated cellulose aerogel for removal of Congo Red from aqueous solution. *Cellulose* **2015**, doi:10.1007/s10570-015-0675-2.

46. Jin, L.; Sun, Q.; Xu, Q.; Xu, Y. Adsorptive removal of anionic dyes from aqueous solutions using microgel based on nanocellulose and polyvinylamine. *Bioresour. Technol.* **2015**, doi:10.1016/j.biortech.2015.08.093.

47. Batmaz, R.; Mohammed, N.; Zaman, M.; Minhas, G.; Berry, R. M.; Tam, K. C. Cellulose nanocrystals as promising adsorbents for the removal of cationic dyes. *Cellulose* **2014**, *21*, 1655–1665, doi:10.1007/s10570-014-0168-8.

48. Yu, H. Y.; Zhang, D. Z.; Lu, F. F.; Yao, J. New approach for single-step extraction of carboxylated cellulose nanocrystals for their use as adsorbents and flocculants. *ACS Sustain. Chem. Eng.* **2016**, *4*, 2632–2643, doi:10.1021/acssuschemeng.6b00126.

49. Manna, S.; Roy, D.; Saha, P.; Gopakumar, D.; Thomas, S. Rapid methylene blue adsorption using modified lignocellulosic materials. *Process Saf. Environ. Prot.* **2017**, *107*, 346–356, doi:10.1016/j.psep.2017.03.008.

50. Pei, A.; Butchosa, N.; Berglund, L. A.; Zhou, Q. Surface quaternized cellulose nanofibrils with high water absorbency and adsorption capacity for anionic dyes. *Soft Matter* **2013**, *9*, 2047, doi:10.1039/c2sm27344f.

51. Gopakumar, D. A.; Pasquini, D.; Henrique, M. A.; De Morais, L. C.; Grohens, Y.; Thomas, S. Meldrum's acid modified cellulose nanofiber-based polyvinylidene fluoride microfiltration membrane for dye water treatment and nanoparticle removal. *ACS Sustain. Chem. Eng.* **2017**, *5*, 2026–2033, doi:10.1021/acssuschemeng.6b02952.

3

Synthesis and evaluation of TiO$_2$/ZnO/MgO/chitosan hydrogel beads for the photocatalytic degradation of organic dye under UV light

Dhanya Arikal[a] and Aparna Kallingal[b*]

[a]Research Scholar, Department of Chemical Engineering, NIT Calicut,
Calicut-673601, India
[b]Assistant Professor, Department of Chemical Engineering, NIT Calicut,
Calicut-673601, India
*Corresponding Author- E-mail: aparnak@nitc.ac.in,
Mobile: +919495669099.

Abstract: Textile dyes are very toxic to human and the environment around. TiO$_2$ nanoparticles have been of great interest in treating these organic effluent dyes as it exhibits good photocatalytic activity. While using TiO$_2$ nanoparticle, there is electron-hole pair recombination occurring, by which, there is less photocatalytic degradation of dye solution taking place. ZnO nanoparticle can be used in combination with TiO$_2$ nanoparticle to form TiO$_2$/ZnO nanocomposite which enhances the photocatalytic degradation efficiency of bare TiO$_2$ nanoparticle. MgO nanoparticle, which acts as a barrier for electron-hole pair recombination can be used along with TiO$_2$/ZnO nanocomposite, to form TiO$_2$/ZnO//MgO nanocomposite. In this research work, TiO$_2$/ZnO/MgO nanocomposite has been immobilized in chitosan beads, which act as the photocatalyst in degrading the dye, where chitosan acts as a support for the nanocomposite. The photocatalyst have been characterized by scanning electron microscope, ultraviolet spectrophotometer, X-ray diffraction, and Fourier-transform infrared radiation. The dye degraded 100% within 75 min with TiO$_2$/ZnO/MgO nanocomposite and within 90 min with TiO$_2$/ZnO/MgO nanocomposite immobilized in chitosan beads. While using TiO$_2$/ZnO/MgO nanocomposite in powdered form requires separation of the nanocomposite from the treated solution, while using immobilization technique, the requirement of separation can be avoided.

Keywords: TiO$_2$/ZnO/MgO nanocomposite encapsulated chitosan beads; photocatalytic degradation; band gap energy; organic dye; absorbance; degradation efficiency.

3.1   Introduction

The aggressive growth of human population and also the strengthening of activities in the agricultural and industrial areas have led to a continual increase in the earth's limited supply of freshwater. Therefore, conservation of natural water resources and blooming of new technologies for water and wastewater treatment became one of the essential environmental issues of the current century. Dyes are used in different types of industries such as paper, plastic, leather, pharmaceutical, food, cosmetics, dyestuffs, textiles, etc. and considerable amount of colored wastewater is generated from these industries [1]. Synthetic dyes are released to the environment mostly by the textile industries. Taking into consideration both effluent composition and volume of effluent discharged, the waste water released by textile industry is the most polluting among all industry sector. The global consumption of dyes and pigments is 7×10^5 tons/year, two-third of which is being consumed by textile industry alone [2,3]. Dyes are having complex chemical structure, which are very difficult to degrade. Wastewater containing textile dyes are toxic and carcinogenic. More than 50% of the dyes used in textile indus-tries are azo dyes. Azo dyes are having one or more azo bonds (-(-N=N-)-). Removal of dyes from the wastewater has become an important issue of interest during the last few years due to the toxicity and persistence of azo dyes.

3.1.1   Methods for removal of dyes from textile effluents

Many methods have been developed and established over the years to remove color from wastewater streams. The treatment technologies used for the deco-lourization of dyes from industries include microbiological decomposition, enzymatic decomposition, chemical precipitation, adsorption on organic and inorganic supports, flocculation, etc. But complete degradation of dyes can-not be achieved by these techniques. Advanced oxidation processes (AOP) is now found to be one of the most efficient technologies for treatment of dyes in waste water streams[4].

Heterogeneous photocatalysis is a type of advanced oxidation process which oxidizes the organic pollutants present in aqueous solutions [5]. It has been considered as a cost effective alternative for the treatment and purifica-tion of dye containing wastewater. Photocatalysis can be defined as a process by which a semiconducting material absorbs light of energy more than or equal to its band gap energy, generating holes and electrons, which can further react to form free-radicals in the system to oxidize the substrate. The produced free-radicals are very efficient oxidizers of organic matter [6]. The photocata-lyst may be in powder form or fixed on a solid support.

In photocatalytic degradation, the complex structure of the dye degrades into simpler and less toxic compounds and hence is of great importance nowadays. Hence, it controls and degrades the contaminants in the pollutant dye solution to a great extent [7]. Semiconductor photocatalysis has attracted a great deal of attention in recent years. Many semiconductor oxides have been used in photocatalytic reactions. The semiconductor nanoparticles have properties between molecules and bulk solid semiconductors. Their physicochemical properties of nanomaterials are found to be strongly size dependent. The nano- scale materials show interesting physical properties such as increasing semiconductor band gap due to electron confinement [8].

Semiconductor photocatalysts which have been used in the photocatalytic redox reactions include TiO_2, ZnO, WO_3, Fe_2O_3, CdS, ZnS, SnO_2,etc. Among the various semiconductors used in heterogeneous photocatalysis, anatase phase of TiO_2 is considered to be a good photocatalyst due to its high photosensitivity and large band gap. It has a wide bandgap of 3.2 eV, high mechanical strength, high refractive index, and chemical stability [9]. When TiO_2 is irradiated with UV light, electrons and holes are generated which helps in the formation of free radicals, that can oxidize the pollutant compounds efficiently. ZnO is also an appropriate photocatalyst as TiO_2, which has almost the same band gap energy as TiO_2 and exhibits high activity in comparison to other existing catalysts [10,13]. The drawback of using TiO_2 nanoparticle is that, when used in photocatalysis, the recombination rate of electron-hole pair is high in the irradiated particles [14]. Low heat capacity, chemical inertness, optical transparency and high thermal stability of MgO nanoparticles make it to be used to reduce the recombination rate of TiO_2 [15]. By combining an insulator (Mg O or Al_2O_3) with TiO_2 the photocatalytic performance can be enhanced [16,17]. And also when ZnO and TiO_2 nanoparticles are combined to form the nanocomposite TiO_2/MgO, the efficiency of the photocatalyst will be increased much more than TiO_2 alone. Hence TiO_2/ZnO/MgO nanocomposite will be of better photocatalytic efficiency towards degrading synthetic dyes. But the ultimate recovery of the photocatalyst is a costly process, when it is used in the slurry form. Immobilization of TiO_2 on different supports started from the early 1980s. Chitosan, which can be manufactured from the natural polymer chitin, can be used as a support for immobilization of TiO_2/ZnO/MgO nanocomposite. By using TiO_2/ZnO/MgO nanocomposite immobilized on chitosan beads there is the combined effect of photodegradation–adsorption. TiO_2/ZnO/MgO nanocomposite degrades the dye molecule into CO_2, H_2O, and mineral salts fast and need a continuous supply of dye molecules otherwise it would undergo recombination. The chitosan which adsorb dye molecules, continuously supplies them for degradation by TiO_2/ZnO/MgO nanocomposite thereby preventing the electron–hole recombination. Here, the objective is to study the degradation efficiency of immobilized

TiO_2/ZnO/MgO nanocomposite in the removal of methyl orange dye in a photocatalytic reactor. Charge transfer and charge separation are the most essential aspects in photocatalytic reaction. Developing TiO_2 based multi–heterogeneous photocatalyst by surface modification has been an effective strategy for charge transfer in enhancing the separation of photogenerated electrons and holes.

3.1.2   Mechanism of photocatalysis

Figure 3.1 describes the mechanism of photocatalysis by TiO_2 [18]. The mechanism of photodegradation process under UV-visible light illumination involves an electron excitation into the conduction band of the TiO_2 semiconductor leading to the generation of very active oxygenated species that attacks the dye molecules leading to photodegradation. That is, when TiO_2 absorbs ultraviolet (UV) radiation from sunlight or illuminated light source (fluorescent lamps), pairs of electrons and holes are produced as in Eq. (3.1). That is, the valence band electron becomes excited when illuminated by light and the excess energy of this excited electron promotes the electron to the conduction band of TiO_2. Hence in TiO_2 negative-electron (e^-) and positive-hole (h^+) is created.

$$TiO_2 + h\nu \rightarrow TiO_2 \, (e^- + h^+) \qquad (3.1)$$

This stage is referred to as the semiconductor's 'photo-excitation' state. The energy difference between the valence band and the conduction band is known as the 'band gap'. The positive-hole breaks apart the water molecule to form hydroxyl radical as in Eq. (3.2).

$$h^+ + H_2O \rightarrow H^+ + OH^\cdot \qquad (3.2)$$

The negative-electron reacts with oxygen molecule to form super oxide anion as in Eq. (3.3). This cycle continues when light is available.

$$e^- + O_2 \rightarrow TiO_2 + O_2^{\cdot -} \qquad (3.3)$$

The hydroxyl radical and super oxide anion reacts with the dye molecule, thus degrading it, to give carbondioxide (CO_2), water (H_2O), and mineral acids as in Eqs. (3.4) and (3.5).

$$^\cdot OH + Dye \rightarrow CO_2 + H_2O + \ldots. \qquad (3.4)$$

$$O_2^{\cdot -} + Dye \rightarrow CO_2 + H_2O + \ldots. \qquad (3.5)$$

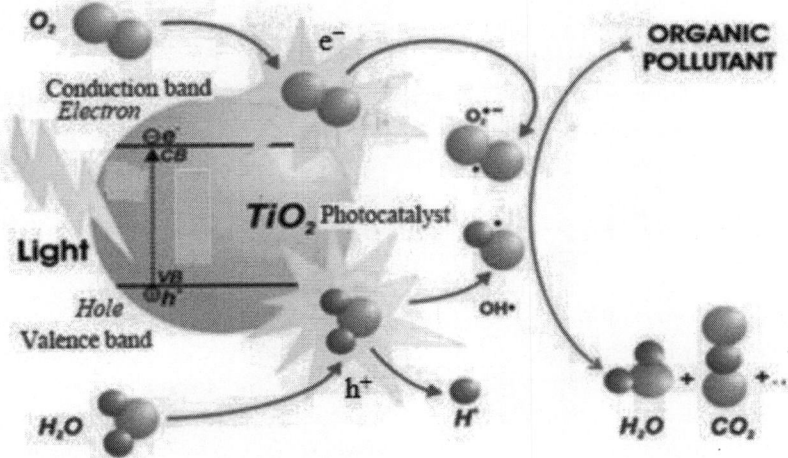

Figure 3.1   Mechanism of photocatalysis

Here, methyl orange (MO) is used as the model dye compound for determining the photocatalytic degradation of TiO_2/ZnO/MgO nanocomposite and TiO_2/ZnO/MgO nanocomposite immobilized in chitosan beads. Methyl orange is an azo dye of basic dye type having azo group (–N = N–) linked to methine or aromatic sp^2 hybridized C atoms[19].

Structure of methyl orange

$$H_3C \diagdown N \diagdown \diagup benzene \diagup N=N \diagup benzene \diagup S \diagup O^- Na^+$$

3.2   Materials and methods

3.2.1   Materials

$TiCl_3$, ammonia solution, magnesium nitrate hexa hydrate, zinc acetate, methanol, sodium hydroxide, chitosan, acetic acid and methyl orange.

3.2.2   Synthesis of the Photocatalyst

Synthesis of TiO_2/ZnO/MgO nanocomposite encapsulated chitosan beads

TiO_2/ZnO/MgO nanocomposite was synthesized from $TiCl_3$. $TiCl_3$ solution was mixed with deionized water and the pH was maintained between 4.5 and

6.5 using ammonia solution. The blue violet colored solution is kept on a magnetic stirrer, which results in a white colored suspension. The white colored suspension is washed with distilled water in order to remove the chloride ions and the precipitate is filtered out. This white coloured precipitate is the TiO_2 nanoparticle [20]. ZnO nanoparticle was prepared from zinc acetate and sodium hydroxide (NaOH)[21]. MgO nanoparticle was prepared from magnesium nitrate hexahydrate $((MgNO_3)_2 \, 6H_2O)$ and NaOH solution [22]. The synthesized TiO_2 nanoparticle, ZnO nanoparticle, and MgO nanoparticle were taken in their respective ratios in distilled water in order to prepare the TiO_2/ZnO/MgO nanocomposite. A well dispersed solution was prepared by vigorous stirring of the nanoparticles followed by ultrasonication for 1h. The dispersion was then heated at 100°C to obtain a dry powder. This dry powder TiO_2/ZnO/MgO nanocomposite is used for further photocatalytic treatment.

TiO_2/ZnO/MgO nanocomposite encapsulated chitosan beads were prepared by dropping the solution of 2wt.% (w/v) chitosan in 2% (v/v) acetic acid solution and TiO_2/ZnO/MgO nanocomposite to 0.5mol/L NaOH solution.

3.2.3   Catalyst characterization

The TiO_2/ZnO/MgO nanocomposite is characterized using scanning electron microscope (SEM), UV (ultraviolet) spectrophotometer, X-ray diffraction (XRD), and fourier transform infrared spectroscopy (FTIR).

3.2.3.1   UV analysis

The UV analysis is done using a UV-vis spectrophotometer. It gives the range of wavelength at which the synthesized nanoparticle can absorb UV radiation. The band gap energy E_g can be determined by applying the Kubelka-Munk function. The K-M function is based on the following equation:

$$F(R) = \frac{(1-R)^2}{2R}$$

where R is the reflectance; $F(R)$ is proportional to the absorption coefficient α.

The indirect band gap of TiO_2 nanoparticle can be determined from the diffuse reflectance spectra (UV – DRS spectra). The band gap is estimated from a plot of $(\alpha h v)^{1/2}$ versus photon energy (hv) by using the relationship.

$$\alpha h v = A \, (hv - E_g)^n$$

where hv is the incident photon energy, E_g is the band gap energy, and A is a constant. The exponent n depends on the type of transition and it may have values 1/2, 2, 3/2, and 3 corresponding to the allowed direct, allowed indirect, forbidden direct, and forbidden indirect transitions, respectively [23]. The value of band gap can be determined by extrapolating the straight line portion of $(\alpha h v)^{\frac{1}{2}}$ on hv axis.

3.2.3.2   Scanning electron microscope(SEM)

A scanning electron microscope is a type of electron microscope that produces images of a sample by scanning the surface with a focused beam of electrons. The electrons interact with atoms in the sample, producing various signals that contain information about the surface topography and composition of the sample. SEM analysis is done to estimate the size and surface morphology of the synthesized TiO_2/ZnO/MgO nanocomposite.

3.2.3.3   Fourier transform infrared radiation (FTIR)

The FTIR spectrum is a graph showing absorbance or transmittance versus frequency or wavelength. The FTIR spectra determines the different bonds present in TiO_2 nanoparticles, ZnO nanoparticles and MgO nanoparticles. And also what happens to the bonds when TiO_2/ZnO/MgO nanocomposite is formed. It also determines the type of bond present in TiO_2/ZnO/MgO nano-composite immobilized in chitosan beads.

3.2.3.4   X-ray diffraction (XRD)

XRD pattern provides a wealth of important information about the arrange-ment and spacing of atoms in a crystalline material. X-ray diffraction peaks are obtained satisfying the Bragg condition, $2dsin\theta = n\lambda$, where, d is the inter-planar distance, θ is the Bragg angle; n is a positive integer and λ is the wave length used. The intensity and position of the peaks give a clear idea of the structure of crystals. The average grain size of a sample can be calculated from the broadening of X-ray diffraction lines [24].

3.2.4   Dye analysis

Methyl orange is used as the model organic dye compound. A standard solution was prepared with a concentration of 5 ppm using distilled water. The dye was analyzed by using a UV–vis spectrophotometer. Absorbance was measured at wavelength (λ_{max}) 464 nm to determine the concentration of methyl orange. Calibration experi-ments were done with different dye concentrations inorder to obtain the unknown concentration from the absorbance in the photocatalytic process.

3.2.5   Photocatalytic study

A photocatalytic reactor having a reactor flask of 500 ml capacity, UV lamp of 125W, magnetic stirrer of 0–800 rpm, outer jacket for water circulation which cools the UV lamp, and an exhaust fan was used for the photocatalytic study of TiO$_2$/ZnO/MgO nanocomposite encapsulated chitosan beads. The beads were dispersed inside the reactor flask within the photocatalytic reactor along with the organic dye solution.

The treated solution was taken out for analysis during certain interval of time. The photocatalytic activity of the nanocomposite can be determined by measuring the absorbance of the initial and final dye solution. The absorbance measured using UV–visible spectrophotometer can be used for determining the concentration of the treated dye solution during each interval of time. The concentration of the treated dye solution can be measured from the calibration curve plotted using absorbance of known concentration of the dye solution. Degradation efficiency can be measured by the equation:

$$\text{Degradation efficiency}(\%) = \frac{Co - Ct}{Co} \times 100$$

where C_o is the initial concentration of the dye solution and C_t is the concentration after treatment with the photocatalyst at time t.

3.3   Results and discussion

3.3.1   Characterization of TiO$_2$/ZnO/MgO nanocomposite encapsulated beads

3.3.1.1   UV analysis and band gap determination

The diffuse reflectance spectrum of the TiO$_2$ nanoparticles is as shown in Fig. 3.2. From the diffuse reflectance spectrum (Fig. 3.2) it can be determined

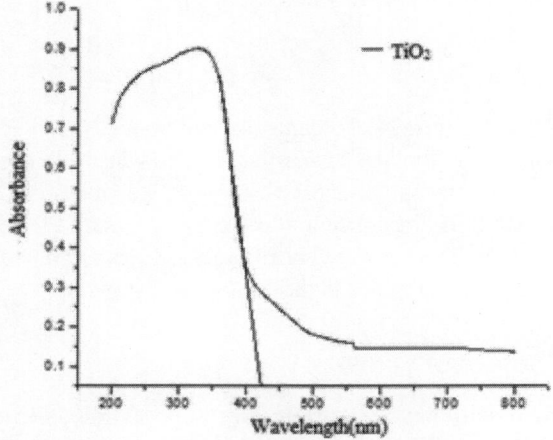

Figure 3.2   Diffuse reflectance spectrum of nanoparticles

that TiO_2 nanoparticles are having absorption around 420 nm. The band gap energy is determined by extrapolating the straight line portion of $(\alpha h v)^{\frac{1}{2}}$ on hv axis; as shown in Fig. 3.3 and it is found that TiO_2 nanoparticles have a band gap of 3 eV.

DRS spectra of TiO_2/ZnO/MgO nanocomposite are as in Fig. 3.4. Maximum absorbance of TiO_2/ZnO/MgO nanocomposite is at 355 nm. The absorption range of the nanocomposite is almost similar with absorption range of TiO_2 nanoparticle. The absorption range of the nanocomposite is

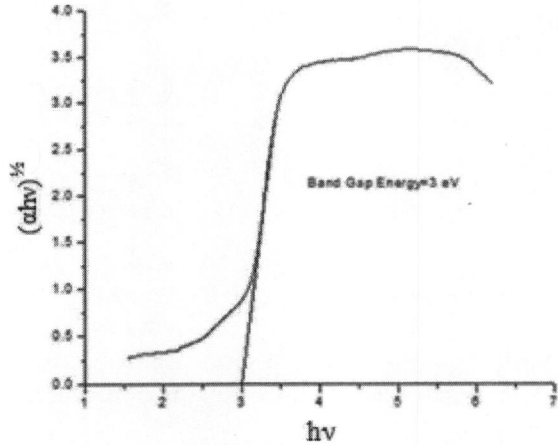

Figure 3.3   Band gap determination of TiO_2 the TiO_2 nanoparticles

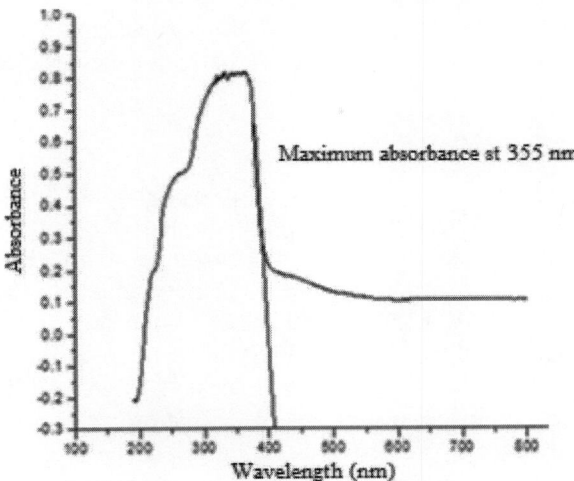

Figure 3.4   Diffuse reflectance spectrum of TiO_2/ZnO/MgO nanocomposite

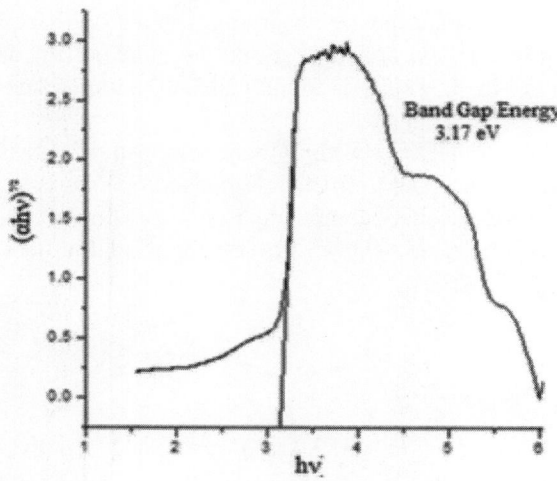

Figure 3.5 Band gap determination of TiO₂/ZnO/MgO nanocomposite

found to be around 405 nm, by which it can be said that, there is not much difference in the band gap energy of TiO_2 with the formation of TiO_2/ZnO/MgO nanocomposite. There is only a slight increase in the band gap energy of the nanocomposite, which may be due to the decrease in the size of the nanoparticles formed, as band gap energy is inversely proportional to the size of the nanoparticles (Fig. 3.5).

3.3.1.2   SEM (scanning electron microscope)

Figure 3.6 shows the SEM image of the synthesized TiO_2/ZnO/MgO nanocomposite. It can be seen that, the size of the particles in the nanocomposite range between 45 nm and 75 nm. The synthesized nanoparticles were found to be spherical in shape.

3.3.1.3   FTIR

Figure 3.7 represents the FTIR spectra of the synthesized TiO_2 nanoparticles, MgO nanoparticles, TiO_2/MgO nanocomposite, and TiO_2/MgO nanocomposite immobilized in chitosan beads in the range of 300–4000cm⁻¹. In TiO_2 nanoparticles,strong absorption in the frequency region of 400–1000 cm⁻¹ corresponds to Ti–O–Ti bonding and indicates the formation of titanium oxide. The broad band observed at 3220 cm⁻¹ is due to the asymmetrical and symmetrical stretching vibrations of hydroxyl group (–OH) and the band at 1636 cm⁻¹ corresponds to deformative vibration of Ti – OH stretching modes. Peak at 421 cm⁻¹ corresponds to the Ti – O bending mode of TiO_2. The band at 2079 cm⁻¹ and 1900 cm⁻¹ indicates weak bonding vibrations of water molecules.

Figure 3.6   SEM of TiO$_2$/ZnO/MgO nanocomposite

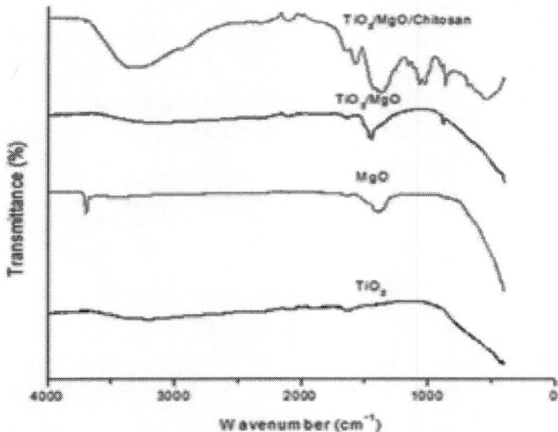

Figure 3.7   FTIR spectra of TiO$_2$ nanoparticles, MgO nanoparticles, TiO$_2$/MgO nano-composites, TiO$_2$/MgO nanocomposite immobilized in chitosan beads

In the spectra of MgO nanoparticles the sharp peak at 3697 cm^{-1} indicates O–H stretching vibration bonds due to water molecules. The peak at 419 cm^{-1} indicates the presence of MgO nanoparticles, which is the strong Mg–O stretching bond. Peak at 834 cm^{-1} were attributed to different Mg–O–Mg vibration modes of MgO. The value of 1640 cm^{-1} represents endothermic peak observed due to very weak bonding vibration of water molecules. The absorption peak at 1477 cm^{-1} is the bending vibration of OH bond, and 1047 cm^{-1} Mg–OH stretching vibrations.

In TiO$_2$/MgO nanocomposite, the absorption peaks were found to be little shifted towards 443 cm^{-1}, which may be the stretching bond of Ti–O–Mg. The value of 1625 cm^{-1}, 3101 cm^{-1}, and 2105 cm^{-1} represents endothermic peak observed due to very weak bonding vibration of water molecules. The shifting of peak from 834 cm^{-1} to 879 cm^{-1} in TiO$_2$/MgO nanocomposite may be due to the vibrational mode of Ti–O–Mg. The value of 1446 cm^{-1} and 1461 cm^{-1} are the bending vibration of OH bond, and 3652 cm^{-1} represents O–H stretching vibration due to the water molecules. Value of 1047 cm^{-1} Mg–OH stretching vibrations were shifted to 1073 cm^{-1} in TiO$_2$/MgO nanocomposite which may be due to the stretching vibration of Ti–O–Mg.

The peak at 443 cm^{-1} in TiO$_2$/MgO and 540 cm^{-1} in chitosan shifted towards 544 cm^{-1} in TiO$_2$/MgO/chitosan indicating the presence of covalent bonding between TiO$_2$/MgO nanocomposite and chitosan polymer. In the same way the peak at 879 cm^{-1} in TiO$_2$/MgO nanocomposite and 894 cm^{-1} in chitosan shifted towards 861 cm^{-1} in TiO$_2$/MgO nanocomposite immobilized in chitosan beads. The other peaks are that of chitosan polymer, that is, the peak at 3365 cm^{-1} indicates the O–H stretching vibration of O–H bond. The peaks in between 1160 cm^{-1} and 1000 cm^{-1} (1148 cm^{-1}, 1110 cm^{-1}, 1066 cm^{-1}, 1028 cm^{-1}) were attributed to vibrations of C–O group. Absorption in the range 1680 – 1480 cm^{-1} (1576 cm^{-1}) was related to the vibrations of carbonyl bonds (C=O) of the amide group.

The FTIR spectra of the TiO$_2$/ZnO/MgO nanocomposite encapsulated chitosan beads are shown in Fig. 3.8. It can be seen that there is the band located at 436 cm^{-1}, which is attributed to the Zn–O stretching mode of the ZnO lattice.

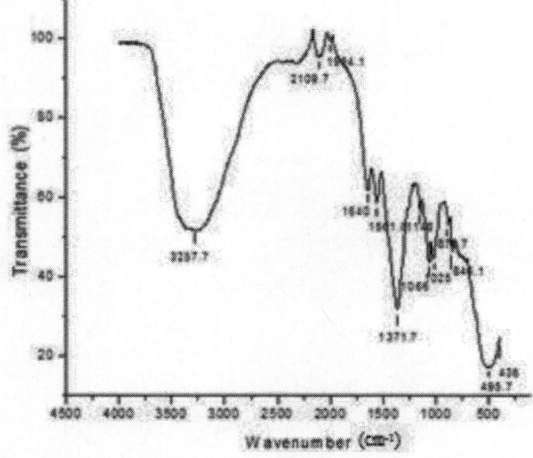

Figure 3.8   FTIR of TiO$_2$/ZnO/MgO nanocomposite encapsulated chitosan beads

3.3.1.4   X-ray diffraction (XRD)

The XRD pattern of the TiO_2/ZnO/ MgO nanocomposite is shown in Fig. 3.9. The peaks corresponding to diffraction angles of 25.3°,38°,48.1°,54.1°,55.1 °,62.8°,68.9°, and 75.2° confirms anatase phase of TiO_2 nanoparticles [25]. Peaks at 31.7°, 34.4°,36°, and 70° corresponds to ZnO nanoparticles and the peak at 63.01° corresponds to MgO nanoparticles. The high intensity XRD peaks indicates that the nanocomposite consisted of nanocrystalline grains.

3.3.2   Degradation experiments

Experiments were conducted to determine the optimum catalyst concentration, optimum pH, and optimum dye concentration for degrading MO dye solution. It was found that percentage degradation increases suddenly, when the cata-lyst concentration increases from 0.05 g/l to 0.1 g/l, and then it is found to be almost stable till 0.5 g/l. Thereafter,when the concentration of the catalyst is increased,the percentage degradation decreases. Increasing the concentration of the catalyst causes an increase in the reaction rate because of an increase of the active site of the catalyst which causes more hydroxyl radical generation; meanwhile, more dye molecules can be adsorbed on the catalyst surface. On the other hand, the additional higher quantities of it would increase opacity of the suspension which in turn would provide reductions in the light intensity throughout the solution. Towards acidic region the degradation efficiency was found to be higher, when compared with the alkaline region. And as the dye

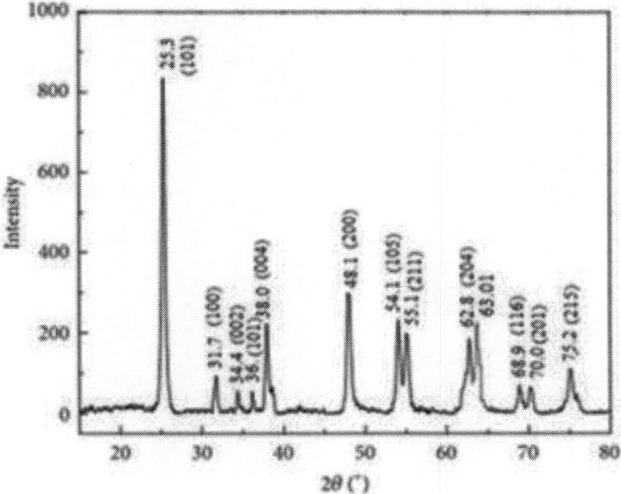

Figure 3.9   XRD of TiO_2/ZnO/MgO nanocomposite

concentration increased, the degradation efficiency decreased as the depth of penetration of UV light decreases which does not make the UV light to reach large portion of the photocatalyst surface.

Therefore, the experimental conditions taken for the degradation experiments were the natural pH of the dye, 0.5 g/L of catalyst concentration and a MO dye solution of 5 ppm. The photocatalytic degradation treatment was done in a photocatalytic reactor of 250 ml Capacity Quartz Reactor, 125 W UV Lamp, spectral range 200 –390 nm UV range and a magnetic drive of 50 – 800 rpm.

After each interval of time, sample dye solutions were taken out of the reactor and analyzed for degradation. By absorbance measurement, the concentration of the sample was found out during each interval of time and the percentage degradation was calculated.

$TiO_2/ZnO/MgO$ nanocomposite could degrade the MO solution completely within 75 min. TiO_2 nanoparticle took 150 min for completely degrading MO dye. Thus it can be concluded that $TiO_2/ZnO/MgO$ is a better photocatalyst when compared to TiO_2 nanoparticles (Fig. 3.10).

$TiO_2/ZnO/MgO$ nanocomposite gave the degradation efficiency of 100% within 75 min towards removal of methyl orange, while $TiO_2/ZnO/MgO$ nanocomposite encapsulated chitosan beads took 90 min for 100% dye removal (Fig. 3.11). The decrease in efficiency in chitosan encapsulated nanocomposite may be due to the scattering of light by the chitosan particles in the nanocomposite.

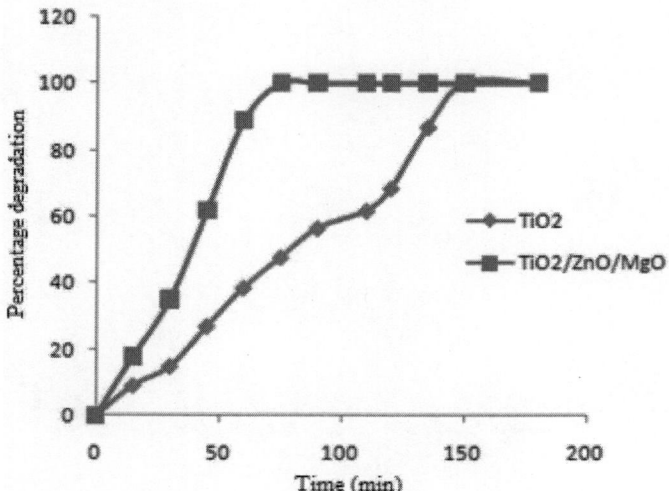

Figure 3.10 Comparison of photocatalytic degradation efficiency of TiO_2 nanoparticles and $TiO_2/ZnO/MgO$ nanocomposite

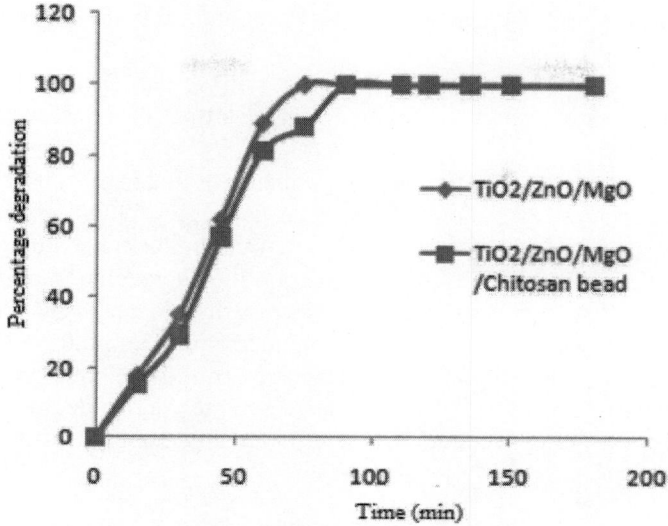

Figure 3.11 Comparison of photocatalytic degradation efficiency of TiO_2/ZnO/MgO nanocomposite and TiO_2/ZnO/MgO nanocompposite encapsulated chitosan beads

3.4   Photocatalytic degradation kinetics

TiO_2/ZnO/MgO nanocomposite and TiO_2/ZnO/MgO nanocomposite encapsulated chitosan beads were observed to be efficient in degrading methyl orange dye solution. The degradation kinetics of methyl orange was described by the Langmuir–Hinshelwood model and followed the first-order kinetics.

Langmuir–Hinshelwood model can be expressed by the following equation:

$$r = -\frac{dC}{dt} = \frac{K_r K_s C_0}{1 + K_s C_0}$$

where, C_0 is the initial concentration of the dye solution, K_r is the reaction rate constant, K_s is the apparent adsorption constant, and t is the time of reaction.

When the concentration is low, $K_s C_0$ is often negligible. The rate of the reaction can then be expressed as first order model as follows:

$$\frac{-dC}{dt} = K_r K_s C_0 = K_{app} C_0$$

where, K_{app} is the apparent first order rate constant

Integration of the equation yields the relation as:

$$\ln\left(\frac{C}{C_0}\right) = -K_{app} t$$

A plot of $-\ln(C/C_0)$ against t will give a straight line if the reaction is of first order.

Kinetics of the photocatalytic reaction with respect to initial dye concentration gives a straight line which indicates that the reaction is of first order.

As the initial concentration of the dye increases, the photocatalytic degradation rate decreases. As shown in Fig. 3.12, using TiO$_2$/ZnO/MgO nanocomposite encapsulated chitosan hydrogel, when the initial concentration of the dye solution increases from 3 ppm to 9 ppm, the degradation rate decreases from 0.035055 min^{-1} to 0.012152 min^{-1}. In case of TiO$_2$/ZnO/MgO nanocomposite in slurry form , the degradation rate decreases from 0.03963 min^{-1} to 0.013599 min^{-1}, when the initial concentration of the dye solution increases from 3 ppm to 9 ppm (Fig. 3.13). The decrease in the rate of degradation with an increase in dye concentration is due to the large amount of dye getting

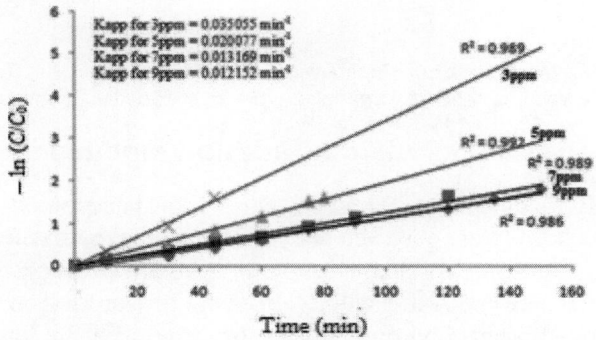

Figure 3.12 Effect of initial dye concentration on degradation efficiency (TiO$_2$/ZnO/ MgO = 0.5 g/L, pH = 8.35)

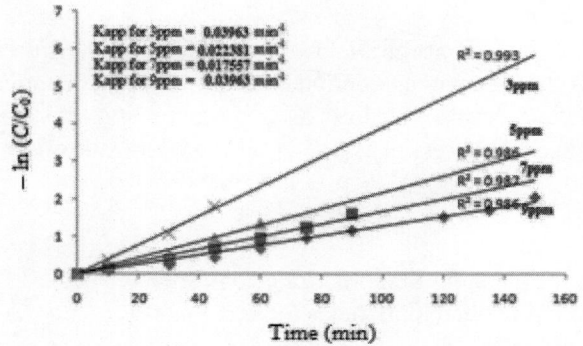

Figure 3.13 Effect of initial dye concentration on degradation efficiency TiO$_2$/ZnO/ MgO nanocomposite

adsorbed on the catalyst surface which results in a decrease in the active sites of the catalyst surface. Thus, there is a decrease in the formation of OH˙ radicals, which is the primary oxidizing agent of the organic dye. And also, with increase in dye concentration,the path length of photons entering the solution decreases according to Beer–Lambert law. Thus, only fewer photons get absorbed on the catalyst surface. And therefore, the production of hydroxyl and superoxide radicals gets limited, which reduces the degradation rate.

3.5   Conclusion

TiO_2/ZnO/MgO nanocomposite is a better photocatalyst compared to bare TiO_2 nanoparticle for the photocatalytic degradation of methyl orange dye solution. Hundred percent degradation of the dyes solution is achieved within 75 min of treatment using TiO_2/ZnO/MgO nanocomposite. Even though the efficiency decreases by using TiO_2/ZnO/MgO nanocomposite immobilized in chitosan beads (complete degradation in 90 min), chitosan acts as a better support for the catalyst and also it decreases the additional cost of treatment while using the unsupported nanoparticles.

References

1. Afkhami, A.; Moosavi, R. Adsorptive removal of Congo red, a carcinogenic textile dye, from aqueous solutions by maghemite nanoparticles. *J. Hazard. Mater.* **2010**, *174*, 398–403.

2. Nigam, P.; Banat, I. M.; Singh, D.; Marchant, R. Microbial Process for the Decolorization of Textile Effluent Containing Azo, Diazo and Reactive Dyes. *Process Biochem.***1996**, *31*(5), 435–442.

3. Robinson, T.; Mcmullan, G.; Marchant, R.; Nigam, P. Remediation of dyes in textile effluent: A critical review on current treatment technologies with a proposed alternative. *Bioresour. Technol.***1996**, *77*(3), 247–255.

4. Sharma, J.; Mishra, I. M.; Kumar, V. Degradation and mineralization of biisphenol A (BPA) in aqueous solution using advances oxidation processes:UV/H_2O_2 and UV/S_2O_8- oxidation systems. *J.Environ. Manage.* **2015**, *156*, 266–275.

5. Augugliaro, V.; Baiocchi, C.; Prevot, A. B.; Garcia-Lopez, E.; Loddo, V.; Malato, S.; Marci, G.; Palmisano, L.; Pazzi, M.; Pramauro, E. Azo-dyes photocatalytic degradation in aqueous suspension of TiO_2 under solar irradiation. *Chemosphere.* **2002**, *49*, 1223–1230.

6. Aarthi, T.; Narahari, P.; Madras, G. Photocatalytic degradation of Azure and Sudan dyes using nano TiO_2. *J. Hazard. Mater.* **2007**, *149* (3), 725–734.

7. Hauas, A. A.; Lachheb, H. A.; Mohamed, K. A.; Elaloui, E. A.; Guillard, C.; Hermann, J. M. Photocatalytic degradation pathway of Methylene blue in water. *Appl. Catal. B.* **2001**, *31*, 145–157.

8. Optical, Electronic, and Dynamic Properties of Semiconductor Nanomaterials, Self-Assembled Nanostructures. In *Nanostructure Science and Technology*; **2004** Springer, Boston, MA, Chapter 8, 201–255.

9. Chen, D.; Caruso, R. A. Recent progress in the synthesis of spherical titania nano structures and their applications. *Adv. Func. Mater.* **2013**, *23*, 1356–1374.

10. Hariharan, C. Photocatalytic degradation of organic contaminants in water by ZnO nanoparticles: Revisited. *Appl. Catal. A.* **2006**, *304*, 55–61.

11. Lizama, C., Freer, J., Baeza, J., Mansilla, H. D. Optimized pho-todegradation of Reactive Blue 19 on TiO$_2$ and ZnO suspensions. *Catal. Today.* **2002**, *76*, 235–246.

12. Poulios, I.; Tsachpinis, I. Photodegradation of the textile dye Reactive Black 5 in the presence of semiconducting oxides. *J. Chem. Technol. Biotechnol.* **1999**, *74*, 349–357.

13. Sakthivel, S., Neppolian, B., Shankar, M. V., Arabindoo, B., Palanichamy, M., Murugesan,V. Solar photocatalytic degradation of azo dye: Comparison of photocatalytic efficiency of ZnO and TiO$_2$. *Sol. Energ. Mat. Sol. C.* **2003**, *77*, 65–82.

14. Ambrus, Z.; Balazs, N.;, Alapi, T.; Wittmann, G.; Sipos, P.; Dombi, A.; Mogyorosi, K. Synthesis, structure and photocatalytic properties of Fe(III)—doped TiO$_2$ prepared from TiCl$_3$. *Appl. Catal. B.* **2007**, *81*, 27–373.

15. Gao, D, Z.; Watkins, M. B.; Shluger, A. L. Transient mobility mechanism of deposited metal atoms on insulating surfaces: Pd on MgO. *J. Phys. Chem.* **2012**, *116*, 14471–14479.

16. Bandara, J.; Kuruppu, S. S.; Pradeep, U. W. The promoting effect of MgO layer in sensitized photodegradation of colorants on TiO$_2$/MgO composite oxide. *Colloids Surf. A: Physicochem. Eng. Asp.* **2006**, *276*, 197–202.

17. Fujishima, M.; Takatori, H.; Tada, H. Interfacial chemical bonding effect on the photocatalytic activity of TiO$_2$/SiO$_2$ nanocoupling systems. *J. Colloids Interface Sci.* **2011**, *361*, 628–631.

18. Karkmaz, M.; Puzenat, E.; Guillard, C.; Herrmann, J. M. Photocatalytic degradation of the alimentary azo dye amaranth Mineralization of the azo group to nitrogen. *Appl. Catal. B.* **2004**, *51*, 183–194.

19. Zainal, Z., Hui, L. K., Hussein, M. Z., Abdullah, A. H., Hamadneh, I. R. Characterisation of TiO$_2$Chitosan/Glass photocatalyst for the removal of a mono-azo dye via photodegradation adsorption process. *J. Hazard. Mater.* **2009**, *164*, 138–145.

20. Cassaignon, S.; Koelsch, M.; Jolivet, J. P. From TiCl$_3$ to TiO$_2$ nanoparticles (anatase, brookite and rutile): Thermohydrolysis and oxidation in aqueous medium, *J. Phys. Chem. Solids.* **2007**, *68*(5–6), 695–700.

21. Osman, D. A. M.; Mustafa, M. A. Synthesis and Characterization of Zinc Oxide Nanoparticles using Zinc Acetate Dihydrate and Sodium Hydroxide. *J. Nanosci. Nanoeng.* **2015**, *1*(4), 248–251.

22. Fedorov, P. P.; Tkachenko, E. A.; Kuznetsov, S. V.; Voronov, V. V.; Lavrishchev, S. V. Preparation of MgO Nanoparticles. *Inorg. Mater.* **2007**, *43*(5), 502–504.

23. Soniya, S. R.; Manikantan Nair, V. Synthesis and Characterization of Nanostructured Mg(OH)$_2$ and MgO. *Int. J. Sci.Res.* **2013**, 197–203.

24. Pankove, J. I. In *Optical Process in Semiconductors*; Prentice-Hall, New Jersey

25. Yuan, W.; Ji, J.; Fu, J. A facile method to construct hybrid multilayered films as a strong and multifunctional antibacterial coating. *J. Biomed. Mater. Res. B Appl. Biomater.* **2007**, *85*(2), 556–563.

4

Different Modifications of Titanium Dioxide Nanoparticles as Photocatalyst in degrading Organic Dyes: A review

Dhanya Arikal[a] and Aparna Kallingal[b,*]

[a]Research Scholar, Department of Chemical Engineering, NIT Calicut,
Calicut-673601, India
[b]Assistant Professor, Department of Chemical Engineering, NIT Calicut,
Calicut-673601, India
[*]Corresponding Author. E-mail: aparnak@nitc.ac.in

Abstract: Synthetic dyes are mostly used in different industries like textile, paper, etc. Dyes are one of the most toxic compound, which pollute the water around. The dyeing and finishing industries are the major sources of pollutants in the industrial sector. Due to the toxicity of these dyes, their removal from textile wastewater has become an issue of interest during the last decade. The conventional methods used for the treatment of these dyes cannot be used for completely mineralizing the organic dye solution. Nowadays, photocatalytic degradation is found to be the most efficient one for degrading such organic dye compounds. Heterogeneous photocatalytic process employs the near UV irradiation to photo-excite a semiconductor photocatalyst. Using semiconductor based photocatalysis, organic contaminants can be totally mineralized, reacting with the oxidizers to produce carbon dioxide, water, and dilute concentration of simple mineral acids. Among the various semiconductors used as photocatalyst, the anatase phase of TiO_2 is known to be a good photocatalyst for the degradation of several pollutants due to its high photosensitivity and large bandgap. When used in nanoscale, TiO_2 is more efficient than the bulk TiO_2 due to its large surface area to volume ratio. The parameters on which the degradation process depends are the intensity of irradiation, dye concentration, catalyst concentration, pH of the dye solution, temperature of the reaction, etc. In this review article, a focus on the different ways in which TiO_2 nanoparticles have been used and also the parameters effecting degradation of organic compounds is being carried out.

4.1   Introduction

Synthetic dyes are the major industrial pollutants and water contaminants [1, 2]. The effluent from the industries which make their way to the environment, pose severe threat to living beings. The discharged effluent have used and unused dyes in it, which makes it a very toxic material for aquatic life and human beings [3]. The release of colored waste waters into the environment causes pollution and eutrophication and will release dangerous byproducts through oxidation, hydrolysis, or through other chemical reactions [4]. Synthetic dyes such as azo dyes have been used in more than 50 % of the dyeing in the textile industry and are the major source of pollutants in the wastewater. Dyes, mostly azo dyes with aromatic structures, are recalcitrant in nature. The azo dyes are difficult to degrade [7]. Wastewater from the dyeing processes is at high temperature and different pH, holding a high amount of the color elements. These concerns have directed to new and/or stringent rules regarding wastewater discharge from dyeing industry and require more efficient treatment methods.

One of the most active areas of environmental research is the development of highly efficient methods. Photocatalytic degradation of textile water effluent is a purification process in which dye is removed from the water by action of semiconductors under a source of light [5]. The process of photocatalytic degradation involves use of semiconductors which absorbs light energy more than or equal to its band gap, generating holes, and electrons which generates free radicals to oxidize the substrate molecule. In photocatalytic degradation, organic compounds decompose by interacting with a photo-catalyst material and UV–vis light and finally CO_2 and H_2O are released [6]. The dominance of photocatalytic degradation using semiconductors in wastewater treatment is because of its advantages over traditional techniques, such as quick oxidation and no formation of polycyclic products. It is an effective and rapid technique for the removal of pollutants from wastewater [7, 8]. In photocatalytic reactions, surface area and number of active sites in the catalyst for absorbing pollutants, play an important role in degrading the dye pollutant. Among catalysts like ZrO_2, ZnO, and TiO_2, high efficiency of TiO_2 has been approved and confirmed in many studies [9]. An ideal photocatalyst should possess the following properties: (i) photoactivity, (ii) biological and chemical inertness, (iii) stability towards photocorrosion, (iv) suitability towards visible or near UV light, (v) low cost, and (vi) lack of toxicity [10].

4.2 TiO_2 photocatalyst

Titanium dioxide is the most widely used photocatalyst nowadays, because of its suitable flat band potential, optical electronic properties, non-toxicity, chemical stability, low cost, and high photocatalytic activity. Titanium dioxide has different crystalline forms. The most common forms are anatase and rutile. The third crystalline form is brookite, which is uncommon and unstable. Anatase is the most stable form [11] and can be converted to rutile by heating to approximately 700°C [12]. The density of rutile is greater at 4.26 g/ml, while anatase has a density of 3.9 g/ml. In the photocatalysis applications, it is known that, anatase is more efficient than rutile, having an open structure compared with rutile.

TiO_2 nanoparticle is one of the best known photocatalyst for degradation of organic contaminants [6]. TiO_2 nanoparticles showed better photocatalytic activity compared to the nano sized TiO_2, because of its large surface area compared to the bulk one. When treating a reactive dye, C.I. Reactive Red 2, the nano- size TiO_2 coated on the polyethylene terephthalate (PET) plastic showed better photocatalytic activity compared to the bulk TiO_2 coated on PET plastic with dye concentration of 100 mg/l. 0.4 g/ml of bulk TiO_2, gave the color degradation of 88% and COD removal of 46%. Furthermore, 0.4 g/ml of nano-size TiO_2, showed enhancement of color degradation and COD removal, that is 98% and 56%, respectively [13].

The anatase phase nano titania (TiO_2) is reported to have better photocatalytic activity compared to the commercial Degussa P-25 catalyst [14, 15]. When using TiO_2 nanoparticles in the degradation of dyes, the photodegradation rate was found to be higher in solvents with higher polarity [16]. Titanate nanoflowers show higher photocatalytic activity for the dye degradation when comparing with titanate nanotubes and nanowires [17].

4.3   Principle of TiO_2 photocatalysis

The proposed mechanism for photocatalytic degradation using TiO_2 nanoparticles for the degradation of dye effluents can be summarized as below [18].

When TiO_2 suspensions are irradiated, electrons are excited from the valence band to the conduction band, generating holes and electrons as in Eq. 4.1:

$$TiO_2 + hv \rightarrow TiO_2\ (e^- + h^+) \tag{4.1}$$

These electrons and holes can recombine as in Eq.4.2.

$$TiO_2\ (e^- + h^+) \rightarrow TiO_2 + heat \tag{4.2}$$

Or can react with some other compounds, for example, oxygen forming superoxide anions which are very reactive as in Eq. 4.3.

$$TiO_2\ (e^-) + O_2 \rightarrow TiO_2 + O_2^{\cdot -} \tag{4.3}$$

The positive holes formed, react with electron donors in the dye solution forming hydroxyl radicals as per Eqs. 4.4 and 4.5.

$$TiO_2 (h^+) + H_2O(ads) \rightarrow TiO_2 + H^+ + OH(ads)^\bullet \qquad (4.4)$$

$$TiO_2 (h^+) + OH^- \rightarrow TiO_2 + OH(ads)^\bullet \qquad (4.5)$$

or it directly oxidize organic matter (OM) at the semi-conductor surface as in Eq. 4.6.

$$TiO_2 (h^+) + OM \text{ degradation intermediates} \rightarrow TiO_2 + OM^{\bullet +} \qquad (4.6)$$

The hydroxyl radicals formed are strong oxidants and can react with the organic matter adsorbed onto the catalyst very quickly, forming oxidized intermediates. Complete mineralization can be achieved, if the treatment is for the required time, which can be represented as in Eq. 4.7.

$$OM + HO(ads)^\bullet \rightarrow \text{degradation intermediates} \rightarrow CO_2 + H_2O + salts \quad (4.7)$$

The reaction mechanism of photocatalysis is characterized by two processes: one chemical and the other physical. The chemical process is a chemical reaction of degradation itself, while the physical process comprises transport to the interface where the solid processing takes place. Figure 4.1 gives the details of all the processes outlined in the TiO_2 nanoparticle photocatalysis [19].

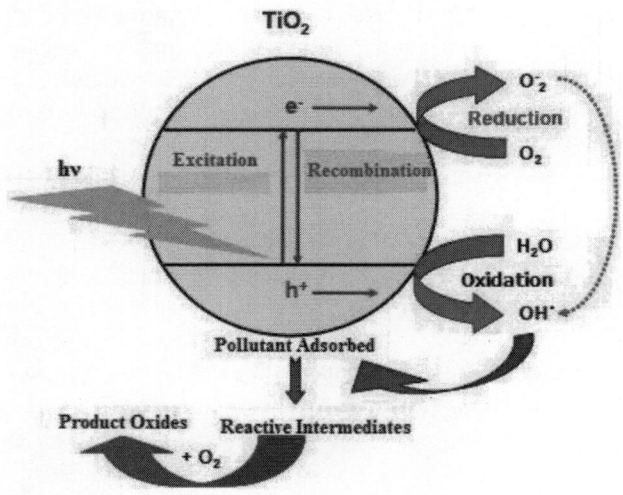

Figure 4.1 Principle of photocatalysis [19]

4.4   Parameters effecting photocatalytic degradation process

There are different parameters on which the photocatalytic degradation of dyes using the TiO_2 nanoparticle photocatalyst and modified TiO_2 nanoparticle photocatalyst depends. They are as follows:

4.4.1   Catalyst concentration

TiO_2 dosage is an important parameter that affects the degradation rate. The initial reaction rates were found to be directly proportional to catalyst concentration indicating the heterogeneous regime. However, it was observed that above a certain level of concentration the reaction rate decreases and becomes independent of the catalyst concentration [20]. The optimum TiO_2 concentration range for photocatalytic degradation of dyes in various studies have been reported between 0.055 [21] and 12.5 g/l [22]. Generally, an increase in concentration of TiO_2 increases the number of active sites on the photocatalyst surface, which in turn increases the number of OH˙ radicals. Besides, when the TiO_2 concentration increases higher than the optimum value, the degradation rate declines due to the interference of the light by the suspension [23, 24], that is, the solution becomes turbid and do not allow the UV rays to pass through and hence blocks the UV radiation[3]. Table 4.1 lists some of the photocatalysts and the conditions to show the effect of catalyst concentration.

4.4.2   Dye concentration

Dye concentration is an important factor to be considered in degradation experiments as pollutant concentration is an important parameter in wastewater treatment. The rate of degradation relates to the probability of formation of OH˙ radicals on the catalyst surface and to the probability of reaction between OH˙ radicals and dye molecules. As the initial concentration of the dye increases, the degradation efficiency of the dye decreases. The presumed reason is that at higher dye concentrations the generation of OH˙ radicals on the surface of catalyst are reduced since the active sites are covered by dye ions. Another possible cause for such results may be the UV-screening effect of the dye itself. At a high dye concentration, a significant amount of UV may be absorbed by the dye molecules rather than the TiO_2 particles which reduces the efficiency of the catalytic reaction because the concentrations of OH˙ and $O_2^{˙-}$ decreases [28, 29].

Table 4.1   Effect of optimum catalyst concentration

Catalyst	Dye	Radiation	Conditions	Optimum catalyst concentration	References
Ag/TiO$_2$ nanophotocatalyst	Methylene blue	UV: 25 W	Dye: 10 ppm pH: 11	0.3947 g/l	[25]
TiO$_2$ nanoparticle	Methylene blue	UV: 25 W	Dye: 10 ppm pH: 11	0.2 g/l	[25]
TiO$_2$ nanoparticle	Acid orange 10	UV: 400 W	Dye: 1×10^{-5} M pH: 3	0.2 g/l	[20]
Fe doped TiO$_2$	Malachite green	Sunlight 32,000–130,000 lux	Dye: 100 mg/l pH: 8.5	0.1 g/l	[26]
TiO$_2$ nanoparticle	Methylene blue	UV: 2.8 mW/cm^2	Dye: 60 mg/l pH: 7	0.9 g/l	[9]
AC/TiO$_2$ nanocomposite	Turquoise blue dye	UV: 2.5 mW/cm^2	Dye: 5mg/l pH: 3	3 mg/l	[27]
5-Sulfosalicylic acid –TiO$_2$ nanoparticle	Methyl orange				

4.4.3   pH of the medium

Considering photocatalytic degradation to be a surface phenomenon, the pH of dye solution is an important parameter in the photocatalytic reaction taking place on the surface of TiO_2 particles as it was revealed by the charge on the particles surface and the size of the nanoparticle. As the zero point charge pH (pHzpc) is about pH 6 for TiO_2, at acidic conditions the particle surface is positively charged and at basic condition, it is negatively charged [30]. If methylene blue is taken as the dye solution, the better efficiency is under alkaline conditions due to the generation of hydroxyl radicals in the alkaline medium. The positively charged TiO_2 absorbs negatively charged methylene blue dye solution and hence degradation efficiency is increased [25]. Table 4.2 lists some of the examples of photocatalysts to show the effect of pH of the medium.

4.4.4   Temperature

The rate of the photocatalytic degradation is temperature dependent and is favored with the increase in solution temperature [33]. In some cases, the temperature effect on degradation yield is found to be negligible. Thus, in those cases, 25°C was introduced as optimum temperature [34].

4.4.5   Calcination

Increasing calcination temperature results in reduced surface area and increased crystallinity and particle size. The catalysts that went through cyclic calcinations (two cycles) at 300°C for a total calcination time of 3.6 h are better than the one subjected to the same calcination time and temperature in a straight run [35].

4.4.6   Light intensity

The light intensity determines the extent of light absorption by the semiconductor photocatalyst at a given wavelength. Light intensity distribution within the reactor invariably determines the overall pollutant conversion and degradation efficiency [36]. In some cases, the rate of reaction exhibited a square root dependency on the light intensity, others observed a linear relationship [37]. It has been reported that the rate is proportional to the radiant flux Φ for $\Phi < 25$ mW/cm^2, and above 25 mW/cm^2, the rate has been shown to be varied as $\Phi 1/2$, indicating a too-high value of the flux and an increase of the electron–hole recombination rate [38]. Initiation of photocatalysis and the formation of electron–hole pairs strongly dependents on the intensity of light [39]. The degradation reaction rate of TiO_2 varies for different intensities of light as

Table 4.2   Effect of pH of the medium

Catalyst	Dye	Radiation	Conditions	Optimum pH	References
TiO$_2$ immobilized on paper	Methyl orange	UV	Light intensity: 125 W Dye: 12 mg/dm^3	3	[31]
Anatase TiO$_2$ nanoparticles	Methylene blue	UV	Light intensity: Six lamps with 20 W Catalyst: 0.1 g Dye: 10 mg/l	9	[32]
TiO$_2$ nanoparticles	Methylene blue	UV	Light intensity: 2.8 mW/cm^2 Dye: 60 mg/l Catalyst: 0.9g/l	7	[9]
AC/TiO$_2$ nanocomposite	Turquoise blue dye	UV	Light intensity: 2.5 mW/cm^2 Dye: 15 mg/l Catalyst: 3 mg/l	3	[27]

follows; the reaction rate increases with increasing light intensity in the range of 0– 20 mW/cm^2. The reaction rate decreases at high-intensity light irradiation due to the favoring of more electron–hole recombination. The excessive light intensity promotes more electron–hole recombination thereby causing decrease in the reaction rate [40 – 43].

4.4.7   Inorganic oxidants

The use of inorganic oxidants, such as H_2O_2, ClO_3^-, BrO_3^-, and $S_2O_8^-$ in TiO_2 system increased the quantum efficiencies either by inhibiting electron–hole pair recombination or by offering additional oxygen atom as an electron acceptor to form the superoxide radical ion O_2^- [44–46]. As in the case of the photocatalyst, there is an optimum of H_2O_2 also, beyond which the rate of decolourization decreases.

Photolysis of hydrogen peroxide results in hydroxyl radical (Eq. 4.8), which is a very powerful oxidizer, able to react with inorganic, aliphatic and aromatic organic compounds [20].

$$H_2O_2 + h\nu \rightarrow 2\ ^\cdot OH \qquad (4.8)$$

4.4.8   Air flow rate

With increase in air flow rate, there is an increase in the bubble formation which enhances the photocatalytic degradation rate. But in reactors, after a limit, there is a decrease in degradation with flow rate, as when the air flow rate is increased, the volume fraction of bubbles increases, which decreases the photon efficiency as increased volume fraction of bubbles leads to mixing of bubbles and also there is a hindrance for the UV light absorption by the photocatalyst [47].

4.4.9   Dissolved oxygen

The presence of oxygen in dye solution enhanced the photocatalytic reaction due to preventing the recombination between the oxidized dye radical and the photoinjected electrons. The presence of air bubbles could enhance the photocatalytic reaction by increasing the mass transfer rates between reactants and catalysts and by scavenging photogenerated electrons [47]. Bandara and Kiwi (1999) reported that pure oxygen was more effective than air for the enhancement of decolorization of dye [48].

4.5   Mineralization

The process, by which a toxic organic compound, an organic dye is converted into carbon dioxide and water etc., is defined as mineralization. Mineralization of target pollutant is the most important thing to be done before discharging the pollutant into the ecosystem. During the treatment process, the organic dye compound undergoes degradation and forms many intermediate compounds, which can sometimes be more toxic than the parent compound. Therefore, the complete mineralization of the substrate should be ensured before discharging these treated solutions into the ecosystem. Determination of carbon content of the oxidation product mixture is the way to measure the oxidation progress of a reaction and here, this can be obtained by monitoring the total organic carbon (TOC) content of the treated solutions .

TiO_2 photocatalyst can be successfully used to degrade the hazardous Tropaeoline 000 dye under UV light irradiation [49] and Rhodamine B under solar light irradiation [50]. The optimized parameters for decolorization of Solophenyl Red 3 BL, under UV light irradiation for the pH value range 2 – 6.3 (pH of the isoelectric TiO_2 semiconductor is pH zc =6.3) is TiO_2 concentration 0.5 g/L , temperature 25°C, initial pH value 2 and for the pH value range 6.3 – 12, the values are TiO_2 concentration 1 g/L, temperature 45°C, and initial pH value 12. The interaction between TiO_2 concentration, temperature, and initial pH was the most influencing parameters [30].

It was found that 0.4% of TiO_2 gave the highest degradation rate constant, when the concentration of methyl orange was 4×10^{-5} M and pH 3 under solar light irradiation, it was found to be 2.6683 h^{-1} and complete degradation was found in 5h of irradiation [51]. After treatment the TiO_2 present in the treated solution will be filtered out or made to settle. To settle completely, sometimes it took one day, but when the pH was made 3, the TiO_2 nanoparticles can settle within a few minutes [51].

Bio-inspired synthesis of highly intricate hierarchical structures using biological materials as templates or precursors is one of the interesting areas of green technology [52]. By developing materials through green synthesis to solve various environmental problems. Use of biotemplates for the synthesis of TiO_2 nanomaterials were of considerable interest. Biotemplates like egg shell membrane [53] and green leafy model [54], using $TiCl_3$, which can hydrolyze in atmosphere at room temperature with a medium rate can be used for the synthesis of TiO_2 nanoparticles. Anatase TiO_2 obtained from yeast and albumen is more efficient than rutile TiO_2 obtained from dandelion pollen for the photocatalytic degradation of methylene blue [55]. In TiO_2 synthesis using *Escherichia coli* as biotemplate, enzyme nitrate reductase present in *E.coli* reduces silver ions in silver nitrate solution to metallic silver [56].

4.6   Modifications of TiO_2

The photocatalytic activity depends on the optical spectrum of TiO_2. And as TiO_2 has a wide band gap (3.2 eV) , TiO_2 will be active only in the presence of UV light, which is <10% of solar light [57]. Inorder to increase the photocatalytic activity of TiO_2 in the visible region, many modifications have been done on TiO_2. The photogenerated electrons and holes are liable for recombination, which can lower the efficiency of photocatalytic degradation. Modification of the electronic band structure of TiO_2 by various procedures, such as coupling with a narrow band gap semiconductor, metal ion/nonmetal ion doping, surface sensitization by organic dyes or metal complexes, noble metal deposition, and codoping with two or more foreign ions, is one of the strategy to overcome the large band gap of titania [58]. By modifying the TiO_2 nanoparticles by depositing some metals or compounds, by doping with some metals, and by synthesizing nanocomposites with some metals or compounds can lower the recombination of these photogenerated electrons and holes, by which the degradation efficiency can be increased. Different modifications of TiO_2 nanoparticles include the following points.

4.6.1   Modification using metal

The enhancement in reactivity by using a noble metal to modify the surface properties was first observed for the photoconversion of H_2O to H_2 and O_2 using the Pt/TiO_2 system [59]. The deposition of metals on the surface of TiO_2 will produce traps to capture the photo-induced electrons or holes.

Figure 4.2 is an illustration of the electron capture properties at the Schottky barrier of the metal in contact with a semiconductor surface. The metal covers only a small area of the semiconductor surface and a large

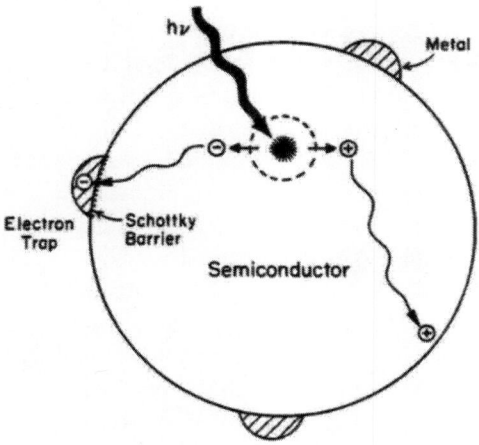

Figure 4.2   Metal modified semiconductor photocatalyst particle [57]

surface area of the semiconductor is still exposed. After excitation the electron migrates to the metal where it gets trapped and electron–hole recombination is suppressed [57].

When metal and semiconductor come in contact the Fermi levels of the two species align causing electrons to flow to the metal from the semiconductor. This leads to an increase in the hydroxyl group because of the decrease in electron density within the semiconductor [60] . Care must be taken when studies are conducted on a metal-modified semiconductor as above the optimum metal content the efficiency of the photocatalytic process actually decreases.

TiO_2 loaded with silver enables the catalyst to perform more efficiently than bare TiO_2 and shortens the illumination period [61] . Photocatalytic activity of $Ag–TiO_2$ is strongly dependent on its phase structure, crystallite size and pore structure. When silver is deposited on TiO_2 nanoparticles, it alters the crystallinity but not the crystal structure, which suggest that the silver deposits are merely placed on the surface of the crystals. The crystallinity gets increased due to the deposition, which is indicated by the increased crystal size with well-developed faces [62]. As anatase TiO_2 is having more photocatalytic activity than rutile phase TiO_2, anatase type $Ag–TiO_2$ has higher photocatalytic activity than rutile type $Ag–TiO_2$ and in anatase phase the samples with small grain size have higher photocatalytic activity than the others [63] . The positive effect of silver on the photoactivity of TiO_2 for the degradation of a Safrani O dye may be attributed to the electronic interaction occurring at the contact region between the metal deposits and the semiconductor surface. Oxygen adsorbed on photocatalyst surface traps the electrons and produces superoxide anion [64]. The holes at the TiO_2 surface can oxidize adsorbed water or hydroxide ions to produces hydroxyl radicals. The $Ag–TiO_2$ catalyst showed 60% increase in degradation efficiency of SO dye as compared to pure TiO_2 with 120 mg/l of the catalyst [62]. When methyl orange dye is used for the degradation process, the degradation is found to be 94.21% for $Ag–TiO_2$ and 63% with bare TiO_2, showing an increase of 33.12% due to the effect of silver doping. 94.21% of degradation was obtained at the optimum catalyst dose of 0.3947g/l at pH 11 with an irradiation time of 80 min [25]. In degrading malachite green of concentration 40 mg/l, $Ag–TiO_2$ of 0.04 g/l with 30.3W/m^2 took 60 min for 100% efficiency [63]. The extent of photocatalyzed mineralization of Acid Red 88 can be obtained within 5 – 7 h of solar light irradiation in the presence of $Ag–TiO_2$. The extent of photocatalyzed mineralization is always higher when peroxomonosulfate (PMS) is used as oxidant than peroxodisulfate (PDS) and H_2O_2. In the case of AR88, only 55% of mineralization was obtained in the absence of oxidants in 7 h but it increased to about 65% (within 6 h) when PMS was used as oxidant. Addition of PDS is also beneficial (58% mineralization in 6 h) for the photocatalyzed mineralization of Acid Red 88 [65].

Furube et al. found that the electron transfer from Au-NPs to TiO_2 is very fast [66]. The photocatalytic activity of mesoporous TiO_2 can be enhanced by Au loading and the 0.25%Au/TiO_2 composite showed the highest photocatalytic activity, which may be ascribed to its high surface hydroxyl content as well as the formed Schottky junction after Au loading [67].

4.6.2   Composite semiconductors

Composite semiconductor photocatalysts provide an interesting way to increase the efficiency of a photocatalytic process. Coupling of semiconductors with appropriate energy levels can produce a more efficient photocatalyst via better charge separation. It also works by extending the energy range of photoexcitation for the system. Coupled semiconductor colloids should be prepared by mixing appropriate amounts of two colloidal suspensions [68]. Figure 4.3 shows charge transfer process taking place when, as an example, CdS – TiO_2 photocatalyst is coupled together. In this CdS–TiO_2 system, photogenerated electron can be transferred from cadmium sulfide into a TiO_2 particle while the holes remain in the CdS particle. The electron transfer from CdS to TiO_2 increases the charge separation and efficiency of the photocatalytic process. The separated electron and hole are then free to undergo electron transfer with adsorbates on the surface [57]. These coupled semiconductor systems have been referred to in the past as "sandwich structure" [68].

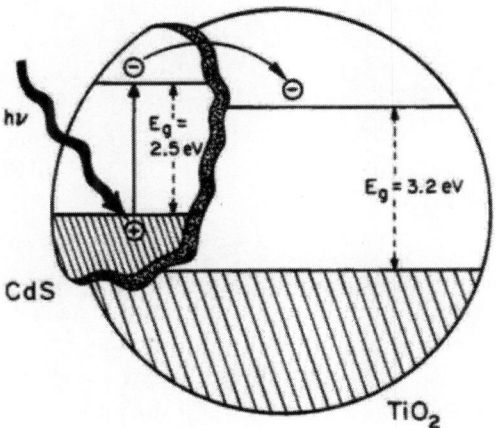

Figure 4.3   Charge transfer processes in a composite semiconductor [57]

4.6.3　Metal oxide semiconductor nanocomposites

Both TiO_2 and ZnO are semiconductors having good photocatalytic activity. The combined effect will be a better photocatalyst compared to both of them individually. The nano-ZnO/TiO_2 improved the separate efficiency of the charge and extended the range of spectrum, which showed a higher efficiency for photocatalytic degradation than the pure nano-TiO_2 and pure nano-ZnO [69]. TiO_2 and ZnO, when used in different molar ratios of 50:0, 50:1, 50:3, 50:5, 50:10, and 50:15, the ratio with 50:10 of TiO_2:ZnO is seen to be the best one, which showed an absorbance value of zero for methylene blue dye for a 2 h irradiation. With increasing the molar ratio of ZnO until 10 the particle size decreased from 72 to 37 nm, but the 50:15 ratios resulted in increasing the particle size until 92 nm. Addition of a high dose of ZnO to TiO_2 decreases the light penetration and reduces the degradation rate [70]. Even though TiO_2 embedded films demonstrated higher photocatalytic efficiency in the deeper UV ranges, the co-integration of TiO_2 and ZnO nanoparticles into the same resin substantially improved the photocatalytic activity in the near-UV and visible spectral ranges, where the intrinsic photocatalytic activities of TiO_2 and ZnO nanoparticles individually were found to be considerably weak [71]. The TiO_2/SiO_2 nanocomposite showed good photocatalytic activity and can be used to completely degrade indigo carmine dye at an optimum catalyst concentration of 3 g/l [72]. When TiO_2 is modified by $CuO–Al_2O_3–ZrO_2$, the bandgap is reduced from 3.25 eV to 1.38 eV, which is probably due to enhanced d-d and charge transfer transitions in the solid phase and is a best choice for degrading environmental dye pollutants [73]. Titanium nano leaves (TNLs) –MgO core – shell has more photocatalytic activity than TNLs, titanium nano particles (TNPs) and the bulk structure [74].

InVO$_4$–TiO_2 nanoparticles can be used to completely mineralize the aqueous solutions of industrial textile azo dyes such as solophenyl red 3BL, coperoxon nevy blue RL, and black nilusun 2BC under visible light and ultrasonic irradiations [34]. SnO_2/TiO_2 nanotube composite showed improved dye degradation efficiency of 85%. The addition of Sn contents with TiO_2 nanotubes show higher degradation at 5% SnO_2 in TiO_2 nanotubes [75].

4.6.3.1　*Magnetic metal oxide semiconductor nanocomposites*

The final separation of photocatalyst from water is one of the disadvantages of using nanoparticles in the degradation process. The special properties of excellent chemical stability, mechanical hardness, cost-effectiveness along with the ability to be separated by a magnet, have made magnetic nanoparticles to be used as an efficient photocatalyst [16]. $Fe_2O_3–TiO_2$ nanocomposite showed super paramagnetic behavior, and it showed better photocatalytic degradation towards methylene blue dye solution [6]. TiO_2/SiO_2-based magnetic nanocomposite ($Fe_3O_4@SiO_2/TiO_2$) under UV irradiation is better than that of $Fe_3O_4@TiO_2$ as the intermediate layer barrier of SiO_2 between the magnetic core and the TiO_2 shell has been proposed to avoid photodissolution of iron and to prevent the magnetic core from acting as an electron–hole recombination centre [76]. TiO_2 based photocatalyst could be separated and reused for three times. For 2 mg of $Fe_3O_4@SiO_2/TiO_2$ nanocomposite, 10 ppm and lower concentration of bromocrosol green showed excellent efficiency [77]. Fe_2O_3 and TiO_2 nanoparticles deposited on Iranian clinoptilolite (Fe_2O_3/TiO_2/clinoptilolite) with great destructive power is synthesized by using Iranian natural zeolite and simultaneous stabilization of nanoparticles of iron and titanium oxides on it is a good photocatalyst inorder to eliminate and photodegrade Acid Black 172, which is one of the most commonly used dyes in textile industries [78].

4.6.4   Nanocomposites with non metals

TiO_2/carbon nanocomposite is an excellent photocatalyst and can be used to degrade reactive dyes, which showed a degradation of 99% of the reactive dye after 60 min. The use of nanocomposites with carbonaceous material can improve the efficiency of the photocatalytic process through a synergistic effect, adsorption of the substrate onto the carbonaceous materials followed by mass transfer to the photoactive TiO_2 [79]. Also, the high electrical conductivity of carbon allows an alternative path for the electron in the conduction band, thus preventing the electron hole recombination [79].

4.6.5   Polymer nanocomposites

The photocatalytic activity of TiO_2/P3HT (Poly(3-hexylthiophene)) can be seen to be better than TiO_2 for the degradation of methyl orange under visible light irradiation, which is a 75% increase. TiO_2/P3HT nanocomposites showed excellent photocatalytic stability after 10 cycles [80]. Wang et al. suggests that P3HT might be a photosensitizer for TiO_2 to decrease the band gap energy of TiO_2 and improve the photocatalytic activity [80].

4.6.6   Doping with non metals

Doping with anion heteroatoms has been extensively pursued as the most potent approach for the development of visible light-activated photocatalysts with nitrogen being the most promising dopant [81]. C–N co-doped TiO_2 nanorods showed improved photocatalytic activity compared to TiO_2 doped with C and N elements under visible light for the degradation of methylene blue. Doping with C and N elements could enhance the corresponding visible-light absorption of TiO_2, and also doping C and N could create more oxygen vacancies in the TiO_2 crystals, which could capture the photogene-rated electrons more effectively, by which more photogenerated holes could be left to improve the photocatalytic activity of TiO_2 [82]. By boron doping, there is an improvement in crystallinity of anatase titania and the increase in amount of superficial hydroxylgroups [83]. While using nanocomposite of activated carbon and nano TiO_2, porosity is an important parameter to con-sider. It is a very good photocatalyst due to its high porosity, super adsorption capability, and low cost [84].

4.6.7   Metal doping in semiconductor

Only 5–6% of day light is ultraviolet radiation, which limits the applicability of TiO_2 as photocatalyst. Doping is one of the techniques to overcome such limitation by which the absorption spectrum extends to infrared radiation i.e., to visible light spectrum (400 – 800 nm), where the day light source could be utilized effectively [26]. During the doping process, the metal ions get integrated into the matrices of the TiO_2. By doping TiO_2 with calcium ions, at a concentration between 0.3 and 1.0 wt.%, its photocatalytic activity can be enhanced, with the best at 0.5 wt.% [35].

4.6.7.1   Transition metal doping

By transition metal doping there is an increase in electron trapping, which inhibits electron–hole recombination. Only certain transition metals such as Fe^{3+} and Cu^{2+} actually inhibit electron–hole recombination [85]. The use of chromium as dopant, create sites which increase the electron–hole recombi-nation [86]. Cerium oxides have attracted much attention due to the optical and catalytic properties associated with the redox pair of Ce^{3+}/Ce^{4+} and the Ce doped TiO_2 showed good photocatalytic activity. When doping with Fe, 0.8% Fe-doped TiO_2 photocatalyst has shown the maximum decolorization of about 98% in degrading crystal violet dye than other pure TiO_2 [87]. With the increase in Sn and Mn content in Sn and Mn doped TiO_2, the average crys-tallite size decreases from 27 to 8 nm with increase in Sn and Mn content in TiO_2. All the Sn doped TiO_2 nanoparticles exhibit anatase – rutile mixed phase while the Mn doped TiO_2 nanoparticles exhibit anatase phase. The optical band gap decreases from 3.24 to 2.21 eV indicates the red shift for Sn and Mn doped anatase and anatase – rutile mixed phase of TiO_2 nanoparticles [88].

In silver doped TiO_2 in the form core–shell nanoparticles with silver core and TiO_2 shell, the core–shell morphology gives several advantages including well defined and higher stability of silver nanoparticles (due to its encapsu-lated state inside TiO_2 shell). The core–shell morphology of silver doped TiO_2 can be obtained only at low Ti:Ag mole ratio (2–6%), with the thicker TiO_2

shell as the higher Ti mole ratio. At high Ti:Ag mole ratio (> 6%), the composite morphology with Ag particles randomly embedded in TiO_2 matrix was formed [89].

The TiO_2 thin films synthesized by doctor blade technique is a fast, non energy consuming process. It produces thin films with high adhesion and high surface porosity, hence good photocatalytic activity. This avoids the problems raised by powder TiO_2, that is leaching and separation. The textile effluents have a very low heavy metal content that can interfere with the photocatalytic degradation process. Cadmium is one among those heavy metals present and it can be adsorbed on the photocatalyst surface, which can modify the process of photocatalytic activity. The modification can be beneficial, if the heavy metal present is doped with TiO_2. Therefore, cadmium precursors used as materials for doping TiO_2 thin films, when used for the degradation of methyl orange and methylene blue, were found to be a better photocatalyst when compared to the bare TiO_2 thin films [90].

4.6.8 TiO_2 on supported materials

When TiO_2 nanoparticles are used for degradation process in suspension, after treatment, the nanoparticles should be removed from the treated water, but their small size does not facilitate their recovery. TiO_2 can be immobilized on various supports like glass, various adsorbents, etc. A suitable support should have a strong attachment to the catalyst, maintain the reactivity of the catalyst, provide a high specific surface area for the reaction and possess high adsorption capacity for the pollutants [5]. When compared to bare TiO_2, immobilization of TiO_2 on supports often reduces the efficiency of the photocatalytic process. The reduction in efficiency occurs as a consequence of reduction in area of the catalyst. These abrasion effects were observed for several authors using alginate beads for different applications [91]. This physical degradation made the beads surface, after the third use, rough with microchannels. Although this phenomenon occurs, the beads can still be used more times, the degradation process is slow and the surface lost is minimal [92].

When TiO_2 particles immobilized on non-woven glass fiber fabric was used for the photocatalytic degradation of reactive dyes, in which decolorization between 21% and 74% can be obtained under solar light irradiation with a COD removal between 0.2 and 0.9g COD/h/m² using TiO_2 concentration of 20 g/m² with a specific area of 250 m²/g [93]. TiO_2 supported on non-woven paper, showed the degradation efficiency of the basic blue dye to be more than 85% within 12 h of irradiation using 10 mg/L of the photocatalyst [94]. Chitosan [β-(1-4)-2-amino-2-deoxy-D-glucose]when used as a binder to anchor the reactive TiO_2 onto the surface of glass plates, there will be the combined effect of photodegradation–adsorption mediated by TiO_2/chitosan. When TiO_2:chitosan changed from 2.5 to 25, the adsorption degradation efficiency increased four times, which is from 47.9% to 87% for methyl orange dye [95], which is attributed to the fact that TiO_2 particles surface are mainly oxygen atoms with a high electron density and can more readily adsorb MO which is a cationic molecule [96]. By taking zeolite as a support for TiO_2 catalyst for degrading Mexican red reactive dye (MRD)(Mixture of C.I. reactive red 2 (50–70%) and C.I reactive orange 86 (15–30%)) and burgundy direct dye (BDD), the degradation was more effective at 1:9 TiO_2 to zeolite ratio with 361 g/l of the initial concentration of the dyes, which gave better results with solar irradiation (96% degradation) than with UV irradiation (92% degradation) with 15 %(w/w) catalyst [5]. TiO_2 coated on nonwoven paper with SiO_2 as a binder can be used to degrade an azo dye Reactive black 5 under UV irradiation [97]. TiO_2 immobilized in a calcium alginate bead retained its photocatalytic activity during all of the experiments and the TiO_2 gel beads presented good stability in water for maintaining its shape after several uses [92]. Table 4.3 lists some of the modifications of TiO_2 used for the degradation of organic dye solutions.

4.6.9   Conversion of visible light to UV light

Visible lights can decompose only some of the azo dye compounds (such as methyl yellow, methyl red, and crystal violet), while the ultraviolet lights can degrade dye compounds completely, which could be an illustration that the holes excited by different wavelength lights possess different abilities to oxidation. By using substances which can transform visible light to ultraviolet light will be a good option to use sunlight for completely degrading the dye solution. The upconversion luminescence materials containing erbium (Er) element mixed into the ordinary rutile TiO_2 powder can be used as such a material for the conversion of visible light to ultra violet light and enhance the photodegradation efficiency [98]. An upconversion luminescence agent, 40 $CdF_2 \cdot 60 \ BaF_2 \cdot 1.0 \ Er_2O_3$, doped TiO_2 can completely degrade ethyl violet within 6 h visible light irradiation, in contrast to 43.24% degradation rate even above 6 h in the presence of undoped TiO_2 [98].

4.7   Recycling of TiO_2 as photocatalyst

Photocatalysis does not involve any waste disposal problem and hence it's a clean technology. The catalyst can be recycled and TiO_2 can be used at least twice without significant change in the efficiency [23]. The economy of the photocatalytic process depends upon how many times a catalyst can be reused

Table 4.3 Modifications of TiO_2 photocatalyst

Catalyst	Dye	Radiation	Conditions	Degradation efficiency (%)	References
Fe_2O_3/TiO_2 nanocomposite	Methylene blue	UV	Dye: 10 ppm Catalyst: 0.04 g/40 ml Light intensity: 400 W	100 (100 min)	[6]
Ag/TiO_2	Methylene blue	UV	Dye: 10 ppm Catalyst: 0.3947 g/l Light intensity: 25 W	94.21 (80 min)	[25]
Fe doped TiO_2	Malachite green	Sunlight	pH: 8.5 Dye: 100 mg/l Catalyst: 0.1 g/l Light intensity: 32,000–130,000 lux	98 (3 h)	[26]
$Fe_3O_4@SiO_2/TiO_2$ nanocomposite	Neutral red bromocresol green	UV	Dye: 10 ppm Catalyst: 2 mg/l pH: 3.7 UV wavelength: 254nm	100(3 h)	[77]
$CuO–Al_2O_3–ZrO_2–TiO_2$ nanocomposite	Direct sky blue 5B	UV	Dye: 100 ppm UV light intensity: 8W Catalyst: 5 mg/l	82.1 (100 min)	[73]
	Direct sky blue 5B	Solar light	Dye: 100 ppm Catalyst: 5 mg/l	96.8 (100 min)	

without sacrificing its efficiency and the type of regeneration it requires. The decrease in the efficiency of the recycled catalyst may be attributed to the deposition of photoinsensitive hydroxides (fouling) on the photocatalysts surface blocking its active sites [20].

4.8   Conclusion

TiO_2 nanoparticle as such is a good photocatalyst. Modifications done in TiO_2 like doping with metals, non metals, etc., using composites with some other semiconductors, compounds, etc., increases the photocatalytic degradation efficiency of the nano photocatalyst. Different dopants show different rates towards the degradation efficiency as different dopants take different positions in the TiO_2 lattice. Even though the modified TiO_2 is a very good photocatalyst, the slurry type reactors used in degradation, becomes a problem in filtering the nanophotocatalyst from the treated solution, which in a way increases the complexity of the reaction and its cost. Therefore, supporting the photocatalyst on a good support will be a good option for this. But the supported photocatalyst shows lower degradation efficiency compared to the slurry type usage of the photocatalyst. Hence, more studies need to be done to increase the efficiency of supported TiO_2 photocatalyst.

References

1. Behnajady, M. A.; Modirshahla, N., Daneshvar, N.; Rabbani, M. Photocatalytic degradation of an azo dye in a tubular continuous-flow photoreactor with immobilized TiO_2 on glass plates, *Chem. Eng. J.* **2007**, *127*, 167–176.

2. Modirshahla, N.; Behnajady, M. A.; Ghanbary, F. Decolorization and mineralization of C . I . Acid Yellow 23 by Fenton and photo-Fenton processes. *Dyes pigments.* **2007**, *73*, 305–310.

3. Akpan, U. G.; Hameed, B. H. Parameters affecting the photocatalytic degradation of dyes using TiO_2-based photocatalysts: A review. *J. Hazard. Mater.*, **2009**, *170* (2–3), 520–529.

4. Giwa, A.; Nkeonye, P. O.; Bello, K. A.; Kolawole, E. G.; Campos, A. M. F. O. Solar photocatalytic degradation of Reactive Yellow 81 and Reactive Violet 1 in aqueous solution containing semiconductor oxides. *Int. J. Appl. Sci. Technol.* **2012**, *2*(4), 90–105.

5. Ochieng, A.; Netshitangani, P.; Akach, J.; Onyango, M. S.; Africa, S.; Engineering, M. Solar/UV photocatalytic degradation of two commercial textile dyes. *J. Hazard. Mater.* **2012**, *2*(1), 36–48.

6. Abbasi, A.; Ghanbari, D.; Masood, M. S. Photo-degradation of methylene blue : Photocatalyst and magnetic investigation of $Fe_2O_3 - TiO_2$ nanoparticles and nanocomposites. *J. Mater. Sci. Mater. Electron.*, **2016**, *27*(5), 4800–4809.

7. Sobana, N.; Muruganadham, M.; Swaminathan, M. "Nano-Ag Particles doped TiO_2 for efficient photodegradation of direct azo dyes nano-Ag particles doped TiO_2 for efficient. *Mol. Catal.* **2006**, *258*, 124–132..

8. Hoffmann, M. R.; Martin, S. T.; Choi, W.; Bahnemannt, D. W. Environmental applications of semiconductor photocatalysis. *Chem. Rev.* **1995**, *95*(1), 69–96.

9. Ehrampoush, M. H.; Ghaneian, M. T.; Rahimi, S.; Ahmadian, M. Removal of methylene blue dye from textile simulated sample using tubular reactor and TiO_2 / UV-C photocatalytic process. *J. Environ. Health Sci. Eng.* **2011**, *8*(1), 35–40.

10. Bhatkhande, D. S.; Pangarkar, V. G.; Beenackers, A. A. C. M. Photocatalytic degradation for environmental applications – A review. *.J. Chem. Technol. Biotechnol.* **2002**, *116* (September 2001).

11. Cotton, M.; Wilkinson, F. A.; Murillo, G.; Bochmann, C. A. In *Advance Inorganic Chemistry*; **1999**. Wiley, New York.

12. Bickley, R. I.; Gonzalez-carreno, T.; Lee, J. S.; Palmisano, Tilleyd, R. J. D. A structural investigation of titanium dioxide photocatalysts. *J. Solid State Chem.* **1991**, *92*, 178–190.

13. Agustina, T. E.; Arsyad, F. S.; Mikrajuddin, A. Photocatalytic degradation of C.I. Reactive Red 2 by using TiO_2 coated PET plastic under solar irradiation. *Adv. Mater. Res.* **2013**, *789*, 180–188.

14. Nagaveni, K.; Sivalingam, G.; Hegde, M. S.; Madras, G. Solar photocatalytic degradation of dyes : High activity of combustion synthesized nano TiO_2. *Appl. Catal. B Environ.* **2004**, *48*, 83–93.

15. Sivalingam, G.; Nagaveni, K.; Hegde, M. S.; Madras, G. Photocatalytic degradation of various dyes by combustion synthesized nano anatase TiO_2. *Appl. Catal. B Environ.* **2003**, *45*, 23–38.

16. Aarthi, T.; Narahari, P.; Madras, G. Photocatalytic degradation of Azure and Sudan dyes using nano TiO_2. *J. Hazard. Mater.* **2007**, *149*(3), 725–734.

17. Huang, J.; Cao, Y.; Liu, Z.; Deng, Z.; Wang, W. Application of titanate nanoflowers for dye removal: A comparative study with titanate nanotubes and nanowires. *Chem. Eng. J.* **2012**, *191*, 38–44.

18. Karkmaz, M.; Puzenat, E.; Guillard, C.; Herrmann, J. M. Photocatalytic degradation of the alimentary azo dye amaranth: Mineralization of the azo group to nitrogen. *Appl. Catal. B Environ.* **2004**, 51, 183–194.

19. Yasmina, M.; Mourad, K.; Mohammed, S. H.; Khaoula, C. Treatment heterogeneous photocatalysis; factors influencing the photocatalytic degradation by TiO_2. *Energy Procedia*, **2014**, *50*, 559–566.

20. Abo-Farha, S. Photocatalytic degradation of monoazo and diazo dyes in wastewater on nanometer-sized TiO_2. *J. Am. Sci.* **2010**, *6*(11), 130–142.

21. Liu, S.; Yang, J.; Choy, J. Microporous SiO_2 – TiO_2 nanosols pillared montmorillonite for photocatalytic decomposition of methyl orange. *J. Photochem. Photobiol. A Chem.* **2006**, *179*, 75–80.

22. Sun, J.; Wang, X.; Sun, J. Photocatalytic degradation and kinetics of Orange G using nano-sized Sn(IV)/TiO_2/AC photocatalyst. *J. Mol. Catal. A Chem.* **2006**, *260*, 241–246.

23. Chakrabarti, S.; Dutta, B. K. Photocatalytic degradation of model textile dyes in wastewater using ZnO as semiconductor catalyst. *J. Hazard. Mater.* **2004**, *112*(3), 269–278.

24. Konstantinou, I. K.; Albanis, T. A. TiO_2-assisted photocatalytic degradation of azo dyes in aqueous solution: Kinetic and mechanistic investigations: A review. *Appl. Catal. B Environ.* **2004**, *49*(1), 1–14.

25. Abdul, M.; Devadi, H.; Krishna, M.; Murthy, H. N. N. Statistical optimization for photocatalytic degradation of methylene blue by Ag–TiO_2 nanoparticles. **2014**, *5*, 1–10.

26. Narayana, R. L.; Matheswaran, M.; Abd, A.; Saravanan, P. Photocatalytic decolourization of basic green dye by pure and Fe, Co doped TiO_2 under daylight illumination. *Desalination.* **2011**, *269*(1–3), 249–253.

27. Gallardo, S. M. Photocatalytic degradation of turquoise blue dye using immobilized AC/TiO_2 : Optimization of process parameters and pilot. *J. Eng. Sci. Technol.* **2015**, 64–73.

28. Daneshvar, N.; Salari, D.; Khataee, A. R. Photocatalytic degradation of azo dye acid red 14 in water on ZnO as an alternative catalyst to TiO_2. *J. Photochem. Photobiol. A Chem.* **2004**, *162*(2–3), 317–322.

29. Tang, W. Z.; Huren An. UV/TiO_2 photocatalytic oxidation of commercial dyes in aqueous solutions. *Chemosphere.* **1995**, *31*(9), 4157–4170.

30. Akila, B.; Mohammed, B. Optimization of photocatalytic degradation of red solophenyl direct dye in TiO_2 aqueous suspension by application of full factorial experimental design; **2014**, doi:10.1109/IREC.2014.6826911.

31. Barka, N.; Assabbane, A. Photocatalytic degradation of methyl orange with immobilized TiO_2 nanoparticles: Effect of pH an d some inorganic anions. *Phys. Chem.* **2008**, *41*, 85–88.

32. Bubacz, K.; Choina, J.; Dolat, D.; Morawski, A. W. Methylene blue and phenol photocatalytic degradation on nanoparticles of anatase TiO_2. *Polish J. Environ. Stud.* **2010**, *19*(4), 685–691.

33. Barka, N. Photocatalytic degradation of an azo reactive dye , Reactive Yellow 84 , in water using an industrial titanium dioxide coated media. *Arab. J. Chem.* **2010**, *3*(4), 279–283.

34. Dianat, S.; Tangestaninejad, S. Preparation, characterization and photocatalytic properties of $InVO_4$ nanopowder and $InVO_4$ – TiO_2 nanocomposite toward degradation of azo dyes and formaldehyde under visible light and ultrasonic irradiation. *J. Iran. Chem. Soc.* **2013**, 535–544.

35. Akpan, U. G.; Hameed, B. H. Enhancement of the photocatalytic activity of TiO_2 by doping it with calcium ions. *J. Colloid Interface Sci.* **2011**, *357*(1), 168–178.

36. Tad, M.; Adesina, A. A.; Pareek, V.; Chong, S. Light intensity distribution in heterogenous photocatalytic reactors. *Asia-Pac. J. Chem. Eng.* **2008**, *3*, 171–201.

37. Terzian, R.; Serpone, N. Heterogeneous photocatalyzed oxidation of creosote components: Mineralization of xylenols by illuminated TiO_2 in oxygenated aqueous media. *J. Photochem. Photobiol. A Chem.* **1995**, *89*(2), 163–175.

38. Herrmann, J. Heterogeneous photocatalysis : Fundamentals and applications to the removal of various types of aqueous pollutants. *Catal. Today.* **1999**, *53*, 115–129.

39. Cassano, A. E.; Alfano, O. M. Reaction engineering of suspended solid heterogeneous photocatalytic reactors. *Catal. Today*, **2000**, *58*(2), 167–197.

40. Malato, S.; Fernandez-Ibanez, P.; Maldonado, M. I.; Blanco, J.; Gernjak, W. Decontamination and disinfection of water by solar photocatalysis: recent overview and trends. *Catal. Today*. **2009**, *147*(1), 1–59.

41. Reza, K. M.; Kurny, A. S.; Gulshan, F. Parameters affecting the photocatalytic degradation of dyes using TiO_2: A review. *Appl. Water Sci.* **2017**, *7*, 1569–1578.

42. Chatterjee, D.; Dasgupta, S. Visible light induced photocatalytic degradation of organic pollutants. *J. Photochem. Photobiol.* **2005**, *C6*(2–3), 186–205.

43. Akpan, U. G.; Hameed, B. H. Parameters affecting the photocatalytic degradation of dyes using TiO_2-based photocatalysts: A review. *J Hazard Mater.* **2009**, *170*(2–3), 520–529.

44. Martin, S. T.; Lee, A. T.; Hoffmann, M. R. Chemical mechanism of inorganic oxidants in the TiO_2/UV process : Increased rates of degradation of chlorinated hydrocarbons. *Environ. Sci. Technol.* **1995**, *29*(10), 2567–2573.

45. Malato, S.; Blanco, J.; Richter, C.; Braun, B.; Maldonado, M. I. Enhancement of the rate of solar photocatalytic mineralization of organic pollutants by inorganic oxidizing species. *Appl. Catal. B Environ.* **1998,** *17*, 347–356.

46. Watanabe, N.; Horikoshi, S.; Kawabe, H.; Sugie, Y.; Zhao, J.; Hidaka, H. Photodegradation mechanism for bisphenol A at the TiO_2 / H_2O interfaces. *Chemosphere*. **2003**, *52*, 851–859.

47. Nam, W.; Kim, J.; Han, G. Photocatalytic oxidation of methyl orange in a three-phase fluidized bed reactor. *Chemosphere*. **2002**, *47*(9), 1019–1024.

48. Bandara, J.; Kiwi, J. Fast kinetic spectroscopy, decoloration and production of H_2O_2 induced by visible light in oxygenated solutions of the azo dye Orange II. *New J. Chem.* **1999**, *23*(7), 717–724.

49. Gupta, V. K.; Jain, R.; Agarwal, S.; Shrivastava, M. Kinetics of photo-catalytic degradation of hazardous dye Tropaeoline 000 using UV/TiO_2 in a UV reactor. *Colloids Surf. A Physicochem. Eng. Asp.* **2011**, *378*(1–3), 22–26.

50. Amini, M.; Ashrafi, M. Photocatalytic degradation of some organic dyes under solar light irradiation using TiO_2 and ZnO nanoparticles. *Nano. Chem. Res.* **2016**, 1(1), 79–86.

51. Al-Qaradawi, S.; Salman, S. R. Photocatalytic degradation of methyl orange as a model compound. *J. Photochem. Photobiol. A Chem.* **2002**, *148*(1–3), 161–168.

52. Brinda, V. G. B. D. K..; Rajan, J. V. V. P. K. S. Enzyme mediated synthesis of Ag – TiO_2 photocatalyst for visible light degradation of reactive dye from aqueous solution. *J. Sol-Gel Sci. Technol.* **2013**, *68*, 60–66.

53. Qi, L.; Ma, J.; Yang, D. Eggshell membrane templating of hierarchically ordered macroporous networks composed of TiO_2 tubes. *Adv. Mater.* **2002**, *14*(21), 1543–1546.

54. Li, B. X.; Fan, T.; Zhou, H.; Chow, S.; Zhang, W.; Zhang, D.; Guo, Q.; Ogawa, H. Enhanced light-harvesting and photocatalytic properties in morph -TiO_2 from green-leaf biotemplates. *Adv. Funct. Mater.* **2009**, *19*, 45–56.

55. Bao, S.-J.; Lei, C.; Xu, M.-W.; Cai, C.-J.; Jia, D.-Z. Nanomaterials for photocatalytic application. *Nanotechnology.* **2012**, *23*(20) 205601.

56. Anil Kumar, S.; Abyaneh, M. K.; Gosavi, S. W.;Kulkarni, S. K.; Pasricha R; Ahmad, A.; Khan, M. I. Nitrate reductase-mediated synthesis of silver nanoparticles from AgNO₃. *Biotechnol. Lett.* **2007**, *29*, 439–445.

57. Linsebigler, A. L.; Lu, G.; Yates, J. T. Photocatalysis on TiO₂ Surfaces: Principles, mechanisms, and selected results. *Chem. Rev.* **1995**, *95*, 735–758.

58. Kumar, S. G.; Devi, L. G. Review on modified TiO₂ photocatalysis under UV/visible light : Selected results and related mechanisms on interfacial charge carrier transfer dynamics. *J. Phys. Chem. A.* **2011**, *115*, 13211–13241.

59. Damme, V. Volume 72, number 1 Chemical Physics Letters 15. **1980**, 72(1), 2–5.

60. Pichat, N. J. P. Effect of deposited Pt particles on the surface charge of TiO₂ aqueous suspensions by potentiometry, electrophoresis, and labeled ion adsorption. *J. Phys. Chem.* **1986**, (13), 2733–2738.

61. Sig, M.; Hong, S.; Mohseni, M. Synthesis of photocatalytic nanosized TiO₂ – Ag particles with sol – gel method using reduction agent. *J. Mol. Catal. A Chem.* **2005**, *242*, 135–140.

62. El-kemary, M.; Abdel-moneam, Y.; Madkour, M.; El-mehasseb, I. Enhanced photo-catalytic degradation of Safranin-O by heterogeneous nanoparticles for environmental applications. *J. Lumin.* **2011**, *131*(4), 570–576.

63. Activity, P. Preparation and photocatalytic properties of silver doped titanium dioxide nanoparticles and using artificial neural network for modeling of photocatalytic activity. *Aust. J. Basic Appl. Sci.* **2011**, *5*(12), 2889–2898.

64. Coleman, H. Effects of Ag and Pt on photocatalytic degradation of endocrine disrupting chemicals in water. *Chem. Eng. J.* **2016**, *113*(October), 65–72.

65. Anandan, S.; Sathish Kumar, P.; Pugazhenthiran, N.; Madhavan, J.; Maruthamuthu, P. Effect of loaded silver nanoparticles on TiO₂ for photocatalytic degradation of Acid Red 88. *Sol. Energy Mater. Sol. Cells.* **2008**, *92*(8), 929–937.

66. Du, L.; Furube, A.; Yamamoto, K.; Hara, K.; Katoh, R.; Tachiya, M. Plasmon-induced charge separation and recombination dynamics in gold - TiO₂ nanoparticle systems : Dependence on TiO₂ particle size. *J. Phys. Chem. C.* **2009**, *113*, 6454–6462.

67. Wang, H.; Luís, J.; Dong, S.; Chang, Y. Mesoporous Au/TiO₂ composites preparation, characterization , and photocatalytic properties. *Mater. Sci. Eng. B*, **2012**, *177*(11), 913–919.

68. Gopidas, K. R.; Bohorquez, M. Photophysical and photochemical aspects of coupled semiconductors. charge-transfer processes in colloidal CdS–TiO, and CdS–Ag I Systems. *J. Phys. Chem.* **1990**, *12*, 6435–6440.

69. Zhang, Z.; Yuan, Y.; Fang, Y.; Liang, L.; Ding, H.; Jin, L. Preparation of photocatalytic nano-ZnO /TiO₂ film and application for determination of chemical oxygen demand. *Talanta.* **2007**, *73*, 523–528.

70. Moradi, S.; Aberoomand-azar, P.; Raeis-farshid, S. The effect of different molar ratios of ZnO on characterization and photocatalytic activity of TiO₂/ZnO nanocomposite. *J. Saudi Chem. Soc.* **2016**, 20(4), 373–378.

71. Perkgoz, N. K.; Toru, R. S.; Unal, E.; Sefunc, M. A.; Tek, S.; Mutlugun, E.; Soganci, I. M.; Celiker, H.; Celiker, G.; Demir, H. V. Photocatalytic hybrid nanocomposites of metal oxide nanoparticles enhanced towards the visible spectral range. *Appl. Catal. B Environ.* **2011**, *105*(1–2), 77–85.

72. Alaoui, O. T.; Nguyen, Q. T.; Rhlalou, T. Preparation and characterization of a new TiO_2/SiO_2 composite catalyst for photocatalytic degradation of indigo carmin. *Environ. Chem. Lett.* **2009**, 7(2), 175–181.

73. Ali, Z.; Hussain, S. T.; Chaudhry, M. N.; Batool, S. A.; Mahmood, T. Novel nano photocatalyst for the degradation of sky blue 5b textile dye. *Int. J. Phys. Sci.* **2013**, 8(22), 1201–1208.

74. Ananthajothi, P.; Venkatachalam, P. Synthesis , structural , optical and morphological studies of TiO_2 nanoleaves- MgO core/shell structure and its photocatalytic activity. *IRJET.* **2015**, *2*, 785–791.

75. Rajkumar, K.; Vairaselvi, P.; Saravanan, P.; Vinod, V. T. P.; Cern, M. Visible-light-driven SnO_2/TiO_2 nanotube nanocomposite for textile effluent degradation. *RSC Adv.* **2015**, 20424–20431.

76. Álvarez, P. M.; Jaramillo, J.; López-pi, F.; Plucinski, P. K. Preparation and characterization of magnetic TiO_2 nanoparticles and their utilization for the degradation of emerging pollutants in water. *Appl. Catal. B Environ.* **2010**, *100*, 338–345.

77. Ahangar, L. E.; Movassaghi, K.; Emadi, M.; Yaghoobi, F. Photocatalytic application of TiO_2/SiO_2 -based magnetic nanocomposite (Fe_3O_4 @ SiO_2/TiO_2) for reusing of textile wastewater. *Nanochem. Res.* **2016**, *1*(1), 33–39.

78. Arimi, A.; Farhadian, M. Assessment of operating parameters for photocatalytic degradation of a textile dye by Fe_2O_3/TiO_2/clinoptilolite nanocatalyst using Taguchi experimental design. *Res. Chem. Intermed.* **2015**, *42*, 4021–4040.

79. Leary, R.; Westwood, A. Carbonaceous nanomaterials for the enhancement of TiO_2 photocatalysis. *Carbon.* **2011,** *49*, 741–772.

80. Wang, D.; Zhang, J.; Luo, Q.; Li, X.; Duan, Y.; An, J. Characterization and photocatalytic activity of poly(3-hexylthiophene)-modified TiO_2 for degradation of methyl orange under visible light. *J. Hazard. Mater.* **2009**, *169*, 546–550.

81. Likodimos, V.; Han, C.; Pelaez, M.; Kontos, A. G.; Liu, G.; Zhu, D.; Liao, S.; De Cruz, A. A.; Shea, K. O.; Dunlop, P. S. M.; Byrne, J. A.; Dionysiou, D. D.; Falaras, P. Anion-doped TiO_2 nanocatalysts for water purification under visible light. *Ind. Eng. Chem. Res.* **2013**, *52*, 13957–13964. .

82. Li, L.; Lu, J.; Wang, Z.; Yang, L.; Zhou, X.; Han, L. Fabrication of the C–N co-doped rod-like TiO_2 photocatalyst with visible-light responsive photocatalytic activity. *Mater. Res. Bull.* **2012**, *479*(6), 1508–1512.

83. Zheng, J.; Liu, Z.; Liu, X.; Yan, X.; Li, D.; Chu, W. Facile hydrothermal synthesis and characteristics of B-doped TiO_2 hybrid hollow microspheres with higher photocatalytic activity. *J. Alloys Compd.* **2011**, *509*(9), 3771–3776.

84. Wang, X.; Liu, Y.; Hu, Z.; Chen, Y.; Liu, W.; Zhao, G. Degradation of methyl orange by composite photocatalysts nano-TiO_2 immobilized on activated carbons of different porosities. *J. Hazard. Mater.* **2009**, *169*, 1061–1067.

85. Butler, C.; Davis, A. P. Photocatalytic oxidation in aqueous titanium dioxide suspensions: The influence of dissolved transition metals. *J. Photochem. Photobiol. A Chem.* **1993**, *70*, 273–283.

86. Disdier, J.; Pichat, P.; Herrmann, J. –M. Effect of chromium doping on the electrical and catalytic properties of powder titania under UV and visible illumination. *Chem. Phys. Lett.* **1984**, *108*, 618–622.

87. Bethi, B.; Sonawane, S. H.; Rohit, G. S.; Holkar, C. R.; Pinjari, D. V.; Bhanvase, B. A.; Pandit, A. B. Investigation of TiO$_2$ photocatalyst performance for decolorization in the presence of hydrodynamic cavitation as hybrid AOP. *Ultrason. Sonochem.* **2016**, *28*, 150–160.

88. Kumar, A.; Chandra, M.; Kumar, P.; Kumar, M.; Soler, M. A. G.; Agarwal, A. Structural, optical and photoconductivity of Sn and Mn doped TiO$_2$ nanoparticles. *J. Alloys Compd.* **2015**, *622*, 37–47.

89. Angkaew, S.; Limsuwan, P. Preparation of silver-titanium dioxide core–shell (Ag@TiO$_2$) nanoparticles: Effect of Ti–Ag mole ratio. *Procedia Eng.* **2012**, *32*, 649–655.

90. Andronic, L.; Enesca, A.; Vladuta, C.; Duta, A. Photocatalytic activity of cadmium doped TiO$_2$ films for photocatalytic degradation of dyes. *Chem. Eng. J.* **2009**, *152*(1), 64–71.

91. Li, Q.; Su, H.; Tan, T. Synthesis of ion-imprinted chitosan-TiO$_2$ adsorbent and its multi-functional performances. *Biochem. Eng. J.* **2008**, *38*, 212–218.

92. Albarelli, J. Q.; Santos, D. T.; Murphy, S.; Oelgemöller, M. Use of Ca–alginate as a novel support for TiO$_2$ immobilization in methylene blue decolorisation. *Water Sci. Technol.* **2009**, *60*(4), 1081.

93. Alinsafi, A.; Evenou, F.; Abdulkarim, E. M.; Pons, M. N.; Zahraa, O.; Benhammou, A.; Yaacoubi, A.; Nejmeddine, A. Treatment of textile industry wastewater by supported photocatalysis. *Dyes Pigments.* **2007**, *74*(2), 439–445.

94. Khataee, A. R.; Fathinia, M.; Aber, S.; Zarei, M. Optimization of photocatalytic treatment of dye solution on supported TiO$_2$ nanoparticles by central composite design : Intermediates identification. *J. Hazard. Mater.* **2010**, *181*(1–3), 886–897.

95. Zainal, Z.; Hui, L. K.; Hussein, M. Z.; Abdullah, A. H.; Moh. I.; Hamadneh, D. K. R. Characterization of TiO$_2$-chitosan/glass photocatalyst for the removal of a mono-azo dye via photodegradation-adsorption process. *J. Hazard. Mater.* **2009**, *164*(1), 138–145.

96. Baran, W.; Makowski, A.; Wardas, W. The effect of UV radiation absorption of cationic and anionic dye solutions on their photocatalytic degradation in the presence TiO$_2$. *Dyes Pigments.* **2008**, *76*, 226–230.

97. Aguedach, A.; Brosillon, S.; Morvan, J.; Lhadi, E. K. Influence of ionic strength in the adsorption and during photocatalysis of reactive black 5 azo dye on TiO$_2$ coated on non woven paper with SiO$_2$ as a binder. *J. Hazard. Mater.* **2008**, *150*(2), 250–256.

98. Wang, J.; Zhang, G.; Zhang, Z.; Zhang, X.; Zhao, G.; Wen, F.; Pan, Z.; Li, Y.; Zhang, P.; Kang, P. Investigation on photocatalytic degradation of ethyl violet dyestuff using visible light in the presence of ordinary rutile TiO$_2$ catalyst doped with upconversion luminescence agent. *Water Res.* **2006**, *40*(11), 2143–2150.

5

Catalytic treatment of organic pollutants using polysaccharide encapsulated metal oxide nanocomposites in assistance with ultrasound: A green chemistry approach

Jayachandrabal Balachandramohan and
Thirugnanasambandam Sivasankar*

Department of Chemical Engineering, National Institute of Technology
Tiruchirappalli, Tamilnadu, India
*Corresponding Author. E-mail: ssankar@nitt.edu

Abstract: Polysaccharides are natural biopolymers that are discerned as promising substitutes of non-degradable polymers due to their exceptional implicit properties of biocompatibility, biodegradability, low cost, and availability. Polysaccharides are highly diversified in terms of structure and functionality, making them suitable for the reduction and stabilization of metal and metal oxide nanoparticles. Substantial research has been focused towards investigating polysaccharide based metal and metal oxide nanoparticles in improving the desired properties in many applications such as catalysis, sensors, etc. This chapter is mainly focused on the association of polysaccharides with metals/metal oxides. The polysaccharide encapsulated metal and metal oxides nanocomposites were synthesized by a sonochemical method and different characterization techniques such as thermogravimetric analysis, X-ray diffraction, Fourier-transform infrared spectroscopy, field emission scanning electron microscopy, high resolution transmission electron microscopy, and UV—visible spectroscopy are discussed in detail. The structure and functionality of the polysaccharide are of great importance for the interaction between metal cations and the chemical functionality of the carbohydrate molecule thus concedes different approaches in achieving the morphology that can be achieved on the one hand, and the potential sustainable use of these biopolymers (a green approach) on the other hand, these material elaboration processes over a unique way for chemists to prepare functional hybrids metal and metal oxides. Besides, the application of the metal and metal oxide nanocomposite was investigated. The catalytic performance of the polysaccharide encapsulated metal and metal oxide nanocomposite for the treatment of organic pollutant was investigated. The catalytic oxidation and reduction of 4-nitrobenzeneamine by Fe_3O_4—guar gum nanocomposite showed an efficiency of 47% and 98% within 60 min. The catalytic oxidation of methyl orange was performed by zerovalent iron—guar gum nanocomposite which showed an efficiency of 99% in 60 min. This chapter can also provide significant insights into the utilization of polysaccharide based metal and metal oxide nanocomposite for the degradation of various organic pollutants.

Keywords: Metal and metal oxides; polysaccharide based metallic nanoparticles; Green chemistry; nanocomposite; polysaccharide encapsulation; catalytic treatment.

5.1   Introduction

Textiles industry is one of the significant contributors in the world that plays a major role in the economies of many countries. However, wastewater is a major environmental hindrance for the growth of textile industry besides other minor issues such as solid waste and resource waste management. Textile industries consume large volumes of water and chemicals for its different wet processing operations. Due to the increasing demand of textile products, there is an enhancement in the number of textile industries and, simultaneously, its wastewater discharge, making it one of the main sources of severe pollution problems worldwide. The discharged wastewater contains chemicals like acids, alkalis, dyes, hydrogen peroxide, starch, surfactants dispersing agents, and soaps of metals [1]. More than 1,00,000 types of commercially accessible dyes are available and an annual production of 7,00,000–10,00,000 tons of organic dyes has been reported previously [2] and in India itself it is close to 80,000 tonnes [3–5]. Probably, 90% of the total dyes produced will end up in fabrics, while the remaining portion will be used in leather, paper, plastic, rubber, concrete, medicine, and chemical industry [6]. Worldwide, an estimated amount of 2,80,000 tons of textile dyes are discharged as industrial effluent every year [7,8].

Dye molecules comprises of two key components: the chromophores which contains conjugated systems of benzene rings bearing simple unsaturated groups (e.g., $-NO_2$, $-N=N-$, $-C=O$) and the auxochromes which contains polar groups (e.g., $-NH_2$, $-OH$) which impart dyeing properties to these compounds but also render the molecule soluble in water and give enhanced affinity (to attach) toward the fibers. The structural diversity of dyes displays considerable attention and was classified in several ways. Hunger [9] classified dyes based on their chemical structure and their application to the fiber type. Dyes may also be classified on the basis of their solubility: soluble dyes which include acid, mordant, metal complex, direct, basic, and reactive dyes; and insoluble dyes including azoic, sulfur, vat, and disperse dyes. Besides this, either a major azo linkage or an anthraquinone unit also characterizes dyes chemically. Based on their usage, the dyes can be classified as acid dyes, cationic (basic dyes), disperse dyes, direct dyes, reactive dyes, solvent dyes, sulfur dyes, and vat dyes. It deliberately provokes to know that around 10–15% of dyes used in industries are harmful to the environment and are discharged into the water streams [10]. They impart color to water which is visible to human eye and therefore, highly regrettable on aesthetic pollution, eutrophication, and perturbations in aquatic life and other associated problems [11,12]. The reduced light penetration reduces the photosynthetic activity and this affects the symbiotic process [13]. The dyes of aromatic, heterocyclically complicated and stable structures used in textile industries are of greater threat for treatment when present in textile wastewaters [14,15]. Most of the dyes that are released to the environment are resistant to degradation, chemically stable, non-biodegradable and exist as substances that possess toxic and carcinogenic characteristics [16]. Therefore, a proper treatment strategy is required to meet the pollution control requirements.

Several techniques and methodologies have been developed for their removal from industry effluents and other water bodies. The different conventional treatment methods that are practiced are physical (adsorption, ion exchange, and membrane filtration), chemical (chemical precipitation, coagulation and flocculation, chemical oxidation, electrochemical oxidation, photo-oxidation, and advanced oxidation process), and biological methods (aerobic and anaerobic). Despite of these dye removal techniques, not all of them are successful or even suitable to be practiced due to their disadvantages [17,18]. Among them, adsorption is the one of the most commonly used and traditional separation technique that have attracted significant attention due to their high affinity, greater decolorization efficiency, and adsorbent regeneration ability [19]. However, most industries require fast removal rate to sustain increasing pollution capacities, an applications of these adsorbents for industrial scale have been restricted by several problems such as its regeneration and/or dumping, sludge generation, and high price of the adsorbent [20,21]. Therefore, adsorbents should be applied to processes that have low concentrations of pollutants or when the adsorbent has a low cost or can be easily regenerated. While processes like flocculation/coagulation results in the generation of large amounts of sludge and the total dissolved solids in the effluent are further increased. The main drawbacks of this treatment method are low removal efficiency, long detention time, and large quantity of sludge generation which required to be removed either by flotation, settling, filtration, or other physical technology to form sludge which is further treated for reducing its toxicity [22,23]. Moreover, large quantities of chemicals are required for coagulation, flocculation, pH, and conductivity adjustment, makes this process uneconomical [24–27]. Electro coagulation treatment of dyes generates metal and metal hydroxide sludge as the end treatment process leading to corrosion problem and hence, this technique was not economically feasible. The conventional biological treatment methods are less efficient because dyes are stable against biological degradation [28], which results in sludge formation, membrane fouling and incomplete mineralization [29–31]. Membrane filtration as a tertiary/final treatment after biological and/or physical-chemical treatments [32,33] will concentrate the dyes [34,35].

Chemical methods such as chemical oxidation using oxidizing agents like ozone, hydrogen peroxide, and chlorine are used to partially or completely degrade the dyes (generally to lower molecular weight species such as aldehydes, carboxylates, sulfates, and nitrogen). The efficiency of the chemical oxidation is hindered by various factors such as reaction parameters [36], types of salt used [37], high cost of hydrogen peroxide, and excess consumption of chemicals [38]. Moreover, the process is unsafe and economically challenging due to the hazards associated with the transport, handling, and storage of bulk quantities of chemicals [39]. Despite this, the use of chlorine causes unavoidable side reactions, producing organochlorine compounds including toxic trihalomethane, thereby increasing the absorbable organic halogens content of the treated water. Also, the liberation of metals in metal complex dyes may cause corrosion in metallic vessels. Therefore, the need to develop sustainable nanomaterials that are economical and offer both high removal rate and high degradation efficiency is inevitable to treat the textile wastewater.

A significant extent on the use of nanomaterials has been under active research and its advancement has been successfully applied in many fields, such as catalysis [40], medicine [41], sensing [42], and biology [43]. In particular, the application of nanomaterials in wastewater treatment has drawn wide attention due to their small sizes and thus large specific surface areas, strong adsorption capacities and reactivity. The most extensively studied nanomaterials for wastewater treatment include zero-valent metals, metal oxides, carbon nanotubes, and nanocomposites. Metal oxides plays a very important role in diverse fields due to their easy mode of formation and multifunctional behavior, and their properties being substantially altered by the porosity, crystallinity, morphology, doping, particle size and, their arrangement in hierarchical structures from the macro- to the nano- and micro-scale. As an emerging and promising technology, the use of iron oxides nanoparticles for wastewater treatment has been extensively studied due to their simplicity and availability. Magnetic magnetite (Fe_3O_4) and magnetic maghemite (γ-Fe_2O_4) and nonmagnetic hematite (α-Fe_2O_3) are often used as nanoadsorbents. Due to their smaller size, separation, and recovery from contaminated water are great challenges in the wastewater treatment. However, magnetic magnetite (Fe_3O_4) and magnetic maghemite (γ-Fe_2O_4) can be easily separated and recovered from the system with the assistance of an external magnetic field. Metal and metal oxide nanoparticles endeavor very fast kinetics and enhanced sorption capacity due to their high surface to volume ratio. However, metal and metal oxide nanoparticles possess impediment against real-time applications, due to weak mechanical strength, lack of specificity for reactions in complex systems and their propensity to aggregate in extremely high pressure drop in flow-through systems [44]. The functionalized/encapsulated metal and metal oxide nanoparticle by polymers and biopolymers are a completely different class of organic materials that are robust, chemically stable, and amenable to chemical modifications depending on the intended application needs. In order to increase the adsorption efficiency and the reactivity of nanoparticles, iron oxides nanoparticles have been functionalized to tune their reactivity using natural polysaccharides for controlling inorganic crystal nucleation and growth of the nanoparticle for the stabilization of nanocomposite.

Naturally occurring polymers are imperative among which polysaccharides are extensively used as stabilizer, capping and reducing agent due to their easy availability, eco-friendly, and non-toxicity. Intensive exploration and research in the past few decades on polysaccharide based metal nanocomposites have led to the emergence of more diverse potential applications exploiting the functionality of these nanomaterials. An environment friendly and sustainable development policy in textile industry requires development of new technologies to reduce water consumption as well as negative environmental impact of discharged wastewater. Researchers have found significant improvements in textile wastewater treatment efficient when polysaccharide-based metal/metal oxide were used as catalyst in advanced oxidation processes. Advanced oxidation processes (AOPs) are the most promising technology for decolorization and mineralization of pollutants present in textile wastewater.

Advanced oxidation processes (AOP) are the processes in which hydroxyl radicals are produced in adequate amounts that will act as prominent oxidizing agents. These hydroxyl radicals have an oxidation potential of 2.80 V compared to other conventional oxidants such as hydrogen peroxide or potassium permanganate which have an oxidation potential of 1.8 V and 1.7 V, respectively. Hydroxyl radicals are also be able to oxidize majority of the complex organic and inorganic chemicals present in the textile effluent and most dyes with high reaction rate constants [47]. The main reaction mechanism of advanced oxidation processes is that organic contaminants are oxidized to CO_2 and H_2O. AOPs comprises a series of methods including chemical oxidation processes using ozone, combined ozone and peroxide, ultra violet enhanced oxidation such as UV/hydrogen peroxide, UV/ozone, UV/air wet air oxidation, and catalytic wet air oxidation (where air is used as the oxidant). AOPs includes photocatalytic oxidation (use of sun light for activation of semiconductor catalyst), and Fenton chemistry (reaction between Fe^{3+} ions and H_2O_2). Fenton's reagent is a convenient chemical (mostly an iron salt) which promote oxidation of complex organic pollutant (by promoting H_2O_2 decomposition), which are resistant to biological degradation. It has also been shown to be operative in degrading both soluble and insoluble dyes. AOP processes also include cavitation, generated either by means of ultrasonic irradiation termed as acoustic cavitation [48] or via constrictions like orifice, venturi, etc. in the hydraulic devices termed as hydrodynamic cavitation.

Ultrasound process, an AOP, utilizes ultrasound waves in the range of 20–1000 kHz when transmitted through an aqueous solution that causes the formation, growth, and collapse of micro sized bubbles, and the rapid collapse of bubbles creates high temperature (about >5000 K), high pressure (about >20 MPa) and high cooling rate (about 10^{10} K/S) [45,46]. Ultrasonic irradiation speeds up the reaction due to the formation of localized cavities, which acts as a micro reactor. Ultrasonic process has wide application including water/wastewater treatment, nanoparticle synthesis, chemical synthesis, etc. A proper synthetic route in designing nanostructured materials has been a driving force for the development of new methodologies and that can be attained with the utilization of high intensity ultrasound. Ultrasound offers a facile, versatile synthetic tool for nanostructured materials that are nonexistence by conventional methods. When liquids are irradiated with ultrasound, the alternating expansive and compressive acoustic waves creates bubbles (that is, cavities) and makes the bubbles oscillate providing a unique interaction between energy and matter. These extraordinary conditions permit access to a range of chemical reaction space normally not accessible, which allows for the synthesis of a wide variety of unusual nanostructured materials. The microscopic implosion of a collapsing bubble that generates high local turbulence and the release of heat energy in the proximity of collapsing bubble "hot spots" can be exploited to enhance the chemical reaction rates of some processes, due to the increased heat and the formation of free radicals.

The synthesis of nanostructured materials by sonochemistry from volatile or nonvolatile precursors, can be explained through different mechanisms [49]. In the case of volatile precursor (for example, a volatile organometallic compound) it will produce free metal atoms by bond dissociation due to the implosive collapse of bubbles with the formation of high temperature. These atoms in the liquid phase nucleate to form nanoparticles or other nanostructured materials in the presence of suitable templates or stabilizers in the solution. Nonvolatile precursors may still undergo sonochemical reactions, even outside of the collapsing bubbles by undergoing reactions with radicals or other high energy species produced from the sonolysis of vapor molecules inside the collapsing bubbles that then diffuse into the liquid phase to initiate a series of reactions (e.g., reduction of metal cations). The synthesis of polysaccharide encapsulated metal oxide nanocomposite by green chemistry approach using ultrasound has been brought to attention in this chapter. Based on the construction strategies of functional nanocomposite, this chapter critically and comprehensively reviews the emerging polysaccharide encapsulated metal oxide nanocomposites in assistance with ultrasound for the catalytic decontamination of organic pollutants. At the same time, the advantages, physicochemical properties and chemical modifications of polysaccharide nanocomposite are also discussed in view of materials development. Finally, the perspective and current challenges of polysaccharide encapsulated metal oxide nanocomposite in future functional nanomaterials are outlined.

5.2   Polysaccharide

Polysaccharides have been proposed as the first biopolymers to have formed on Earth. Polysaccharides are polymers of simple sugars or polymeric carbohydrate molecule composed of long chain of monosaccharide units bound together by glycosidic linkage and on hydrolysis give the constituent monosaccharides or oligosaccharides (The term "saccharide" is derived from the Greek word sakchar, meaning "sugar or sweetness"). Some polysaccharides are homogenous polymers that contain only one kind of sugar (for example, glycogen), while others are complex heterogeneous polymers that contain 8–10 types of sugar. Polysaccharides are cyclic monomers bound together by glycosidic bonds to form highly complex structures utilized widely for diverse applications.

5.2.1   Classification of polysaccharide

Polysaccharides can be categorized based on their source of origin (natural or synthetic), hydrophilicity (hydrophobic or hydrophilic) and molecular weight (low or high), etc. Polysaccharides classified based on their source of origin was shown in Fig. 5.1. With growing interest towards bio-degradable, renewable, biocompatible and non-toxic material synthesis, more stress has

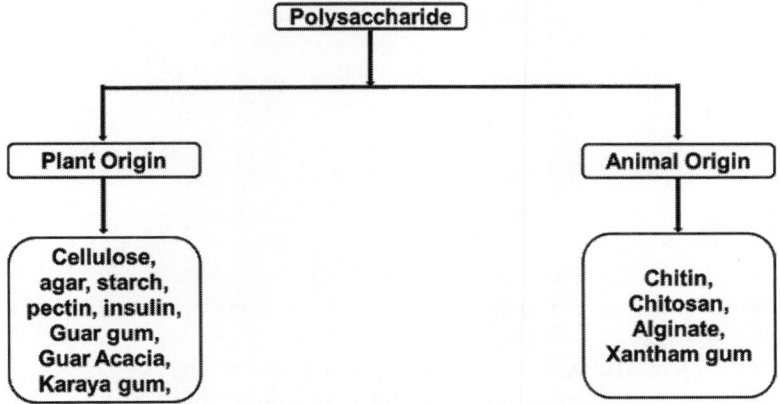

Figure 5.1 Classification of polysaccharide

been laid on the use of natural polysaccharides. In addition to biodegradable nature, naturally occurring polysaccharides are stable, cheap, and easily available.

Among the polysaccharide classification based on their origin from plant and animal sources especially obtained from natural sources, the following polysaccharide have been used for the nanoparticle synthesis.

5.2.2   Polysaccharide used for nanoparticle synthesis

5.2.2.1   *Cellulose*

Cellulose is an organic polysaccharide consisting of a linear chain of several hundred to over ten thousand β-(1$\rightarrow$4) linked D-glucose units having the formula $(C_6H_{10}O_5)_n$ [50]. The plant cell wall mainly consists of cellulose, hemicelluloses, and pectin [51]. He et al. [52] succeeded in the versatile synthesis of noble metal (platinum, palladium, and silver) nanoparticles in colloidal suspension by using a porous cellulose fibers as novel stabilizers. The implication of the results indicates that nanoporous structure and the high oxygen (ether and hydroxyl) density of cellulose fibers anchor metal ions tightly via ion–dipole interactions which were stabilized by strong bonding interaction with their surface metal atoms. Chang et al. [53] and Yu et al. [54] used carboxymethyl cellulose sodium as a stabilizer in the preparation of Sb_2O_3 and ZnO nanoparticles as the filler in glycerol plasticized pea-starch. The tensile strength of Sb_2O_3/carboxymethyl cellulose sodium increases with increase in the concentration from 0% to 5% and decrease in the water vapor permeability. Chang et al. [55] synthesized magnetic Fe_3O_4 nanoparticles by using carboxymethyl cellulose sodium as stabilizers in order to improve the stability, biocompatibility, and biodegradability. Lin et al. [56] synthesized ZnO/Zn nanocomposite by corn starch and cellulose as chelating agents which favored surface area (11.8443–15.7100 m²/g) and pore size (12.3473–13.7453 nm) and the resulting nanocomposite exhibited super catalytic capability for the photodegradation of methylene blue, and Congo red. A novel green thin film of silver nanoparticles was synthesized by Hussain et al. [57] in which hydroxypropylcellulose was used as a template nanoreactor, stabilizer, and capping agent. This furnishes a green and economical strategy for the synthesis and storage of stable Ag NPs. Chook et al. [58] synthesized silver nanoparticles functionalized by cellulose nanofibrils by a green in situ hydrothermal synthesis approach. The functionalized silver nanoparticles aerogel has exhibited a highly porous structure and showed sensitivity for detecting rhodamine B at different concentrations, ranging from 5×10^{-3} M to 5×10^{-7} M. Ghasemzadeh et al. [59] prepared chitosan–carboxymethyl cellulose membrane by cross-linking of chitosan and carboxymethyl cellulose and a novel silver nanocomposite was prepared by using chitosan–carboxymethyl cellulose as a biopolymer matrix. Zhao et al. [60] natural cellulosic fiber matrix and bio-derived polyhydroxyalkano-ate (PHA) were fabricated by dip coating, in which PHA was grafted using maleic anhydride to improve its compatibility with cellulosic fibers. This hydrophobic nanocomposite was used as a substitute for synthetic polymer/cellulose composite materials.

5.2.2.2   Agar

Agar or agar–agar consists of dried gelatinous substance obtained from *Gelidium amansii* (Gelidaceae) and it is also obtained from several other species of red algae like *Gracilaria* (Gracilariaceae), and *Pterocladia* (Gelidaceae) [61]. Agar consists of a mixture of two components, the linear polysaccharide agarose which is made up of the repeating monomeric unit of agarobiose and a heterogeneous mixture of smaller molecules called agaro-pectin, which is a disaccharide made up of D-galactose and 3,6-anhydro-L-galactopyranose. Agaropectin is a mixture of smaller acidic side group molecules with the possible occurrence of sulfate, methoxyl, and/or pyruvate substituents at various positions in the polysaccharide chain [62]. An agar-based silver nanoparticles and nanocomposite film was synthesized using agar extracted from the red alga *Gracilaria dura*. This nanocomposite film had the great bactericidal activity and may find applications in food preserva-tion and wound dressing [63]. Makwana et al. [64] investigated the effect of silver modified montmorillonite prepared by a solution intercalation method using agar–carboxymethyl cellulose bio nanocomposites film for antibac-terial packaging materials for food preservation by controlling foodborne pathogens and spoilage bacteria.

5.2.2.3   Starch

Starch or amylum is the polymeric carbohydrate reserved material in green plants and it is mainly present in seeds and underground organs. Starch occurs in the form of granules (starch grains). A number of starches are recognized for pharmaceutical use and these include maize (*Zea mays*), rice (*Oryza sativa*), wheat (*Triticum aestivum*), and potato (*Solanum tuberosum*) [61]. Starch or amylum is a carbohydrate consisting of a large number of glucose units joined together by glycosidic bonds. It consists of two polymers, namely amylose (a non-branching helical polymer consisting of α-1,4 linked D-glucose mono-mers) and amylopectin (a highly branched polymer consisting of both α-1,4 and α-1,6 linked D-glucose monomers) [65].

Palladium nanoparticles were successfully prepared using mesoporous starch-supported materials as support media by Budarin et al. [66]. The palladium-supported materials were also reusable, preserving their catalytic activities after four reuses and exhibited excellent catalytic activities in the microwave-assisted Heck, Suzuki and Sonogashira C–C coupling reactions. Besides, gold and platinum nanoparticles, starch-based silver nanoparticles (5.8 nm) exhibited a concentration-dependent in hemagglutination and induce significant hemolysis [67]. White et al. [68] reported mesoporous starch as nanoparticle stabilizer, support and reducing surface for the synthesis of silver nanoparticle. The resulting nanocomposite revealed high surface areas ($S_{BET} \geq$ 150 m) and mesopore volumes ($V_{meso} \geq 0.45$ cm^3 g^{-1}) with antimicrobial activity. Valodkar et al. [69] revealed the non-cytotoxic green Cu–starch conjugate offers a rational approach towards antimicrobial application and for integration to biomedical devices. Thirumavalavan et al. [70] compared the photodegradation of methylene blue, crystal violet, and Congo red using effective nano zinc oxide (ZnO) obtained from polysaccharides such as chitosans, corn starch, and sodium alginate as chelating agents. The author revealed that the degradation was 100% for crystal violet whereas for methylene blue and Congo red it was 75% and 57%, respectively, for the photocatalyst obtained from corn starch. Silver nanoparticles are prepared with various environmentally friendly coatings – polysaccharide starch (AgNPs/Starch, Dav=15.4±3.9 nm) and trisaccharide raffinose (AgNPs/Raff, Dav=24.8±6.8 nm) which showed a possible reduction in their reactivity of 2,4-dinitrophenol-stimulated ATPase activity of intact mitochondria [71]. The modified ZnO/Zn nanocomposite by corn starch as chelating agents and the resulting nanocomposite exhibited super catalytic capability for the photo degradation of methylene blue, and Congo red as compared to cellulose [56]. Moreover, Tripathy et al. [72] synthesized biometallic Ag–Au nanoparticles capped by a novel biodegradable graft copolymer hydroxyethyl starch-g-poly (11.1 nm) and exhibited great recyclability (98–93%) after 4 cycles, providing an enhanced efficiency in the 4-nitrophenol reduction reaction and the degradation of 'N–N' (azo bond) bond in some azo dyes in presence of sodium borohydride.

5.2.2.4   Inulin

Inulin are a group of naturally occurring polysaccharides obtained from the bulbs of Dehlia, *Inula helenium* (Compositae), roots of Dendelion, *Taraxacum officinale* (Compositae). Burdock root, *Saussurea lappa* (Compositae) or chicory roots, and *Cichonium intybus* (Compositae) [61]. Inulin is a heterogeneous collection of fructose polymers and consists of chain-terminating glucosyl moieties and a repetitive fructosyl moiety, which are linked by β (2,1) bonds [73]. The degree of polymerization (DP) of standard inulin ranges from 2 to 60. Kalaivani and Suja [74] extracted inulin from *Allium sativum* L. by hot water diffusion method and synthesized a novel bio-nanocomposite by embedding TiO$_2$ (rutile) onto the inulin matrix. The bio-nanocomposite was found to be twice as high as that of pristine TiO$_2$ for the photodegradation of methylene blue.

5.2.2.5   *Guar gum*

Guar gum is a natural water soluble nonionic polysaccharide isolated from the seeds of *Cyamopsis tetragonolobus* [75]. It is also called guaran, clusterbean, Calcutta lucern, Gum cyamposis, and Cyamopsis gum, Guarina, Glucotard, and Guyarem [76]. Chemically, guar gum is a natural polysaccharide consisting of a linear backbone of β-1,4-linked mannose units with α-1,6-linked galactose units as side chains at average galactose-to-mannose ratios of 1:1.6–1:2 [77,78], some of which are linked to single sugar side chains of α-D-galactose [76]. Guar gum has a backbone composed of β-1,4-linked-D-mannopyranoses to which, on average, every alternate mannose an α-D-galactose is linked 1→6 [79].

A green synthesis of silver nanoparticles [80] using guar gum (GG) as a reducing agent and stabilizing agent had revealed that the nanoparticle can be used for the application of aqueous ammonia sensing. Further, the silver nanoparticle dispersion in polymer matrix played a more important role in calorimetric sensing applications. A low cost gold nanoparticles (AuNPs) synthesized using guar gum had displayed the optical sensor property for the detection of aqueous ammonia based on surface plasmon resonance. A detection limit of 1 ppb was possible for both polysaccharide-based gold silver nanoparticle in the sensing of ammonia at room temperature [81]. Pandey and Nanda [82] also reported guar gum-based gold nanoparticle exhibits a wider detection range of ammonia from 0.1 parts-per-quadrillion (ppq) to 75,000 parts-per-million (ppm) due to the variations in electrical conductivity. Rastogi et al. [83] showed the synthesis of palladium nanoparticles immobilized organic–inorganic hybrid nanocomposite have found to exhibit strong electrocatalytic activity towards the oxidation of hydrazine. Das et al. [84] synthesized a highly stable monodispersed silver nanoparticle in the presence of a blend of a polysaccharide (guar gum) which played a vital role in the long-term stabilization of aqueous dispersions of silver nanoparticles. The aqueous dispersions of silver nanoparticles also displayed strong antibacterial and antioxidant properties.

5.3   Synthesis of iron oxide–guar gum nanocomposite for the treatment of 4-nitrobenzeneamine

5.3.1   Synthesis procedure

For the synthesis of iron oxide (Fe_3O_4)–guar gum nanocomposite (IGNC), acid hydrolyzed gaur gum (AHG), and 1 M of $FeSO_4 \cdot 7H_2O$ was subjected to ultrasound irradiation under nitrogen environment. Then, 3 M NaOH was

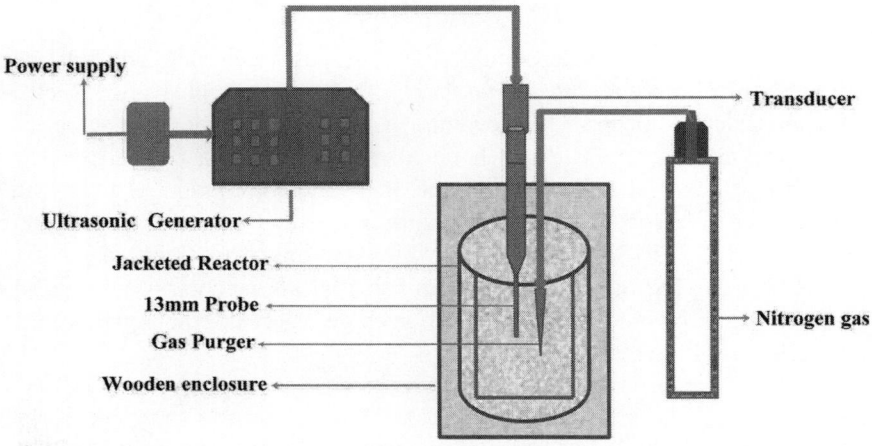

Figure 5.2 A laboratory-scale high intensity ultrasound apparatus

injected in the reaction mixture after 30 min of sonication using the experimental setup as shown in Fig. 5.2. The reaction was halted when the solution turns from blue to black color which indicates the formation of the iron oxide nanocomposites (IGNC). The nanocomposite was washed and dried.

5.3.2   Characterization of iron oxide–guar gum nanocomposite

The TGA curves were observed in the range of 30–900 °C with a heating rate of 10 °C using alumina crucible. The thermal behavior of iron oxide nanocomposite as shown in Fig. 5.3 illustrate the initial mass loss up to 100 °C was due to the hydroxyl radical present in the sample. The thermal gravimetric analysis of AHG, illustrate the mass loss up to 200 °C which was due to the oxidative decomposition of the polymer, vaporization, and elimination of volatile products [87,88]. The thermal decomposition of polysaccharide initiates by random breakdown of glycosidic bonds, followed by further decomposition [89,90]. The gradual mass loss up to 300 °C was observed due to dehydroxylation, but there are no significant thermal effects. The mass gain was observed from 300 °C to 400 °C due to the presence of magnetite based on the exothermic reaction of Fe(II) oxidation to Fe(III). The exothermic reaction with no mass loss, take place around 560 °C which was due to the transition of maghemite (γ-Fe_2O_3) to hematite (α-Fe_2O_3) [85,86]. The thermograms of IGNC reveals that the nanocomposite was more stable than the iron oxide nanoparticles.

Fourier-transform infrared spectroscopy (FTIR) studies were performed for AHG, Fe_3O_4, and IGNC to compare the changes in their chemical structure and the characteristic IR wave number. FTIR spectra of AHG, Fe_3O_4, and

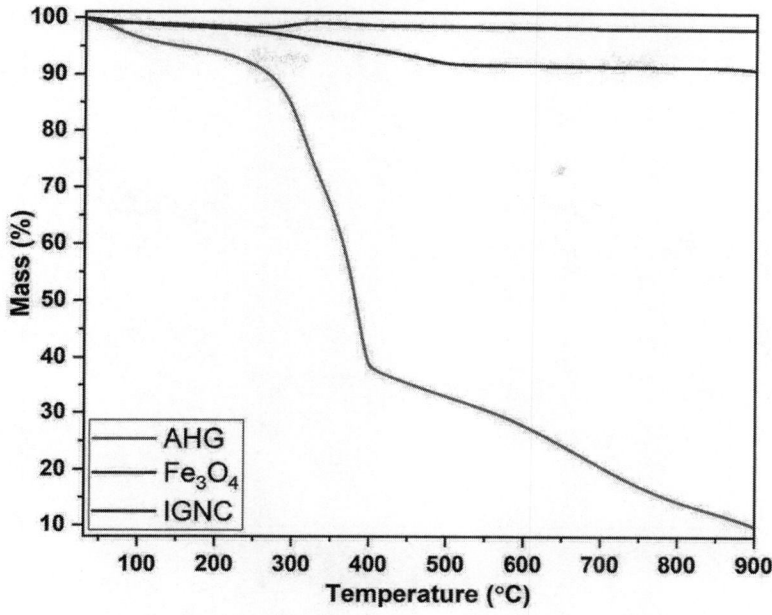

Figure 5.3   TGA thermograms of AHG, Fe_3O_4, and IGNC

IGNC were shown in Fig. 5.4. The spectral regions between 2800 cm^{-1} and 3000 cm^{-1} shows C–H stretching modes [91,92]. The peak observed in the region around 3300 cm^{-1} was due to O–H stretching vibration of polymer and water involved in hydrogen bonding [93]. The strong absorption band present at 3402 cm^{-1} indicated O–H bond stretching vibration. The absorption band due to ring stretching of mannose may be appeared at 1655 cm^{-1}. The absorption bands recorded at 1382 cm^{-1} and 1358 cm^{-1} may due to symmetrical deformations of –CH_2 and –COH groups. The peaks observed in the spectra between 900 cm^{-1} and 1150 cm^{-1} represented the highly coupled C–C–O, C–OH, and C–O–C stretching modes of polymer backbone [92]. Associated water molecule resulted in the band near 1650 cm^{-1} in the spectra. The region around 1400 cm^{-1} due to CH_2 deformation was also observed [94]. The characteristic asymmetrical band region at 500–550 cm^{-1} indicates the Fe–O and Fe–OH stretching vibrations [95]. FTIR spectral regions reveal the functional group of the AHG were observed in IGNC, which confirms that the polysaccharide was present in the nanocomposites.

The crystal structure of the as-synthesized nanoparticles was analyzed by X-ray diffractometer with scans acquisition at 2θ range from $20°$ to $80°$, scan speed of $1°$/min and step size of $0.05°$. The X-ray diffraction patterns of as-synthesized AHG, Fe_3O_4, and IGNC sample were shown in Fig. 5.5.

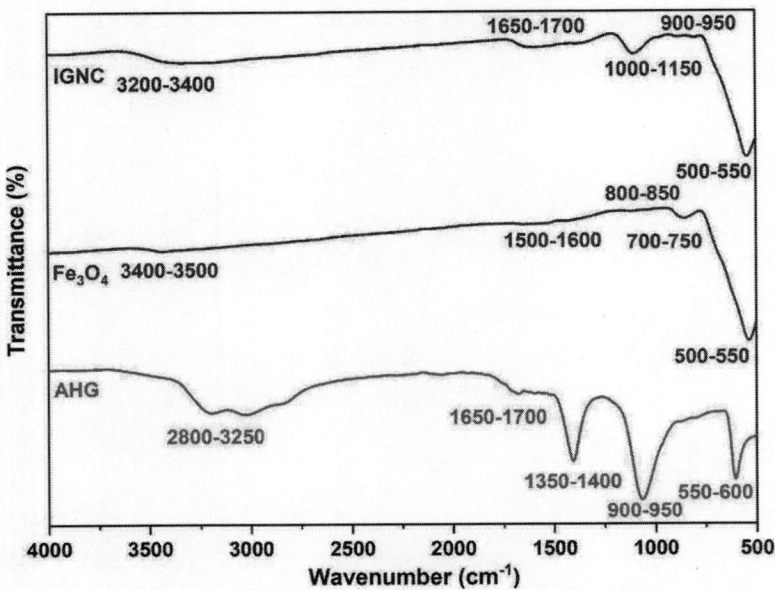

Figure 5.4   FTIR spectra of AHG, Fe_3O_4, and IGNC

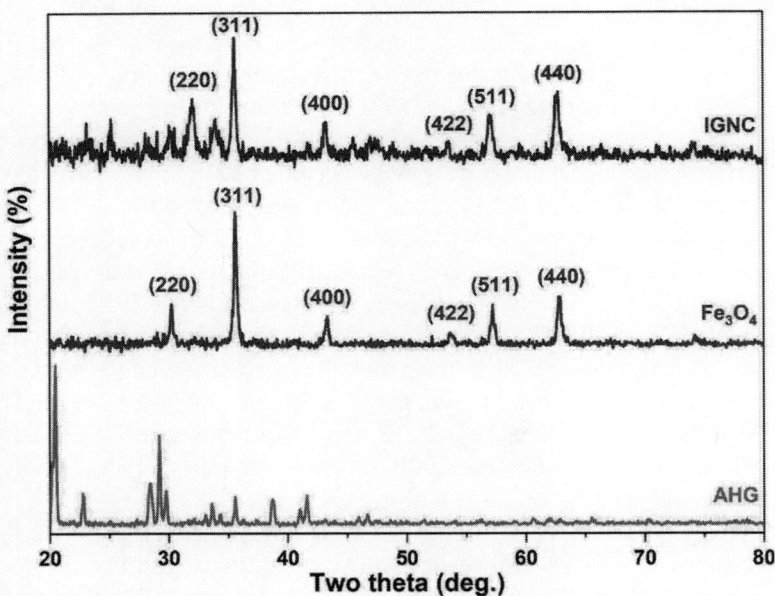

Figure 5.5   XRD patterns of AHG, Fe_3O_4, and IGNC

The peaks can be indexed at 30.1°, 35.4°, 43.0°, 53.39°, 56.9°, and 62.6° corresponding to the crystal planes of (220), (311), (400), (422), (511), and (440) respectively. The sharper and stronger peaks indicate the formation of iron oxide with a cubic inverse spinel structure, which are consistent with the standard data for magnetite (JCPDS card no. 00-019-0629). Results evidenced that guar gum presented amorphous structure. AHG, Fe_3O_4, and IGNC showed overall crystallinity. The IGNC shows the same peaks as those of magnetite nanocubes, indicating that the functionalization does not degrade the core magnetite. However, it is notable that in the case of the IGNC, the peaks height decreased and the background became noisy. This correctly indicates that the peak reflections are originated from the core magnetite and the noisy background comes from the amorphous AHG. Similar results were obtained [96,97] for dextran-coated magnetite and fucan coated magnetite MNPs.

The structural morphology of the magnetite nanocubes and IGNC were characterized using field emission scanning electron microscopy as shown in Fig. 5.6a and b. It reveals that these particles have a relatively narrow size distribution and nearly uniform cube shape. The figure shows that the IGNC improves the dispersion of nanoparticle and the aggregation of particle was

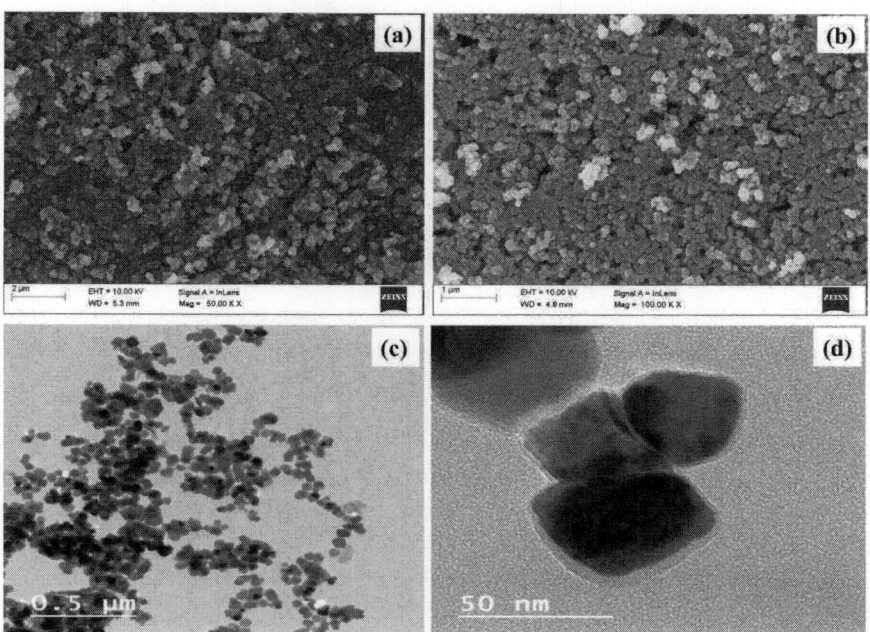

Figure 5.6   (a) SEM image of magnetite nanocubes, (b) SEM image of IGNC nanocomposite, (c and d) TEM images of IGNC nanocomposite

reduced because of the presence of GG that weakens the magnetic interaction between particles. Hence, it is noted that the nanocomposite are of polydispersed single phase magnetite (Fe_3O_4). The high-resolution TEM images were observed to provide the information about the size, shape, aggregation state, as well as crystallinity, and lattice spacing. Figure 5.6c and d shows the low magnification TEM images of the as-prepared IGNC, where the cubical shape with well crystallinity is revealed.

Polysaccharides such as AHG are highly reactive materials with high oxygen content, thus making them promising O-providers. GG is an ideal sustainable substance, certainly "greener" than that of surfactants and additives issued from petrochemical industry, and generally used for the same purpose. Figure 5.6c and d shows that the Fe_3O_4 are encapsulated by the GG polymeric networks or Fe_3O_4–GG core–shell structure. The formed GG polymer networks expedite the excellent stability and the formation of the nanoparticles through electrostatic and steric effects and the hydroxyl groups of the polymeric chains, which further improve the stabilization of IGNC. The polymer GG act as a capping agent and hydroxyl groups of the polymeric chains networks surround and protect the particles for longer periods. On the other hand, a strong physical adsorption of the AHG polymer onto the surface of the Fe_3O_4 nanoparticle is also an indication of better stabilization.

5.3.3   Treatment of *4*-nitrobenzeneamine by Fe_3O_4–guar gum nanocomposites

For the degradation study, *4*-nitrobenzeneamine (NBA) and Fe_3O_4–guar gum nanocomposites (IGNC) was subjected to catalytic oxidation under ultrasound irradiation. The degradation efficiency at different experimental conditions was shown in Fig. 5.7.

The NBA reaction solution was subjected to simple ultrasound with the solution pH 8.0 and optimal pH 3.0, the degradation observed was only 13% and 33%, respectively, whereas in the presence of IGNC catalyst it was around 26% at pH 8 and 47% at pH 3, respectively. The increase in the degradation efficiency was due to the presence of Fe^{2+} ions and the OH^- group in the nanocomposite, and the degradation process proceeds rapidly only at pH 3.0 [98]. The absorbance spectra of the aqueous NBA solution by using IGNC were shown in Fig. 5.8. So, the catalytic oxidation of NBA reveals that only 47% degradation was possible using ultrasound-assisted method, in order to increase the efficiency of the catalyst, catalytic reduction of NBA was performed.

For the catalytic reduction of NBA, $NaBH_4$ a reducing agent was introduced in the NBA reaction solution and was subjected to ultrasound irradiation. The reduction efficiency of NBA was observed as shown in Fig. 5.8.

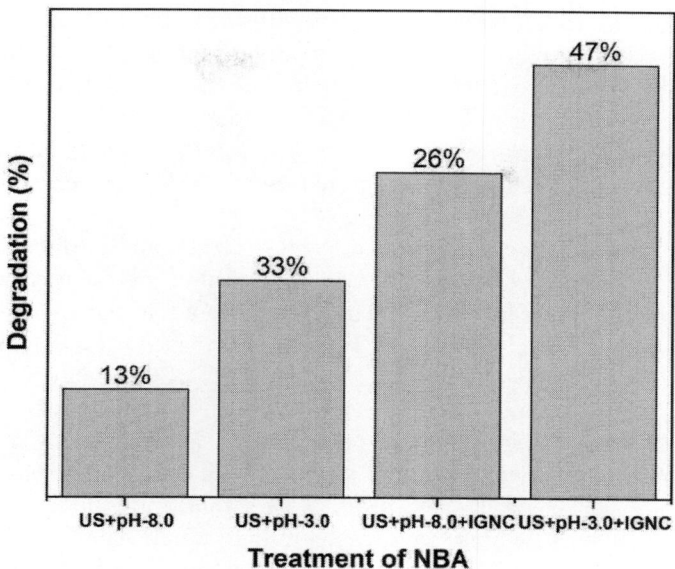

Figure 5.7 Degradation of NBA ($[NBA]_0$=50 ppm, amplitude=40%, probe=25 mm, IGNC=1.0 g/L)

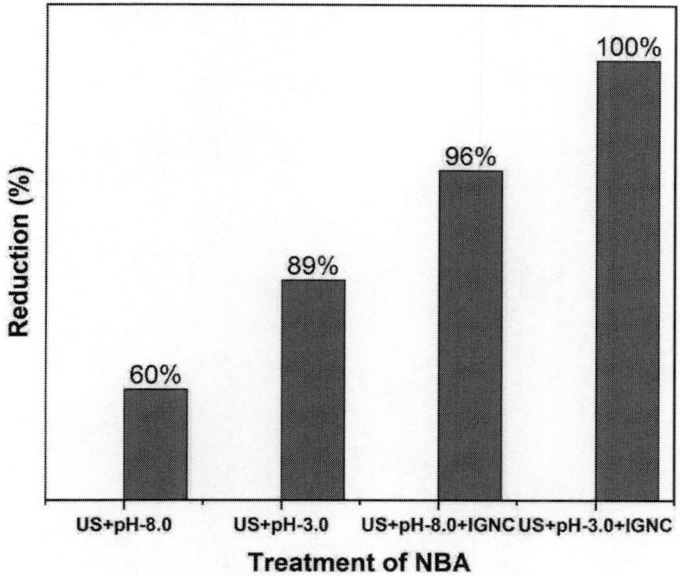

Figure 5.8 Reduction of NBA ($[NBA]_0$, 50 ppm; amplitude, 40%; probe, 25 mm; IGNC, 1.0 g/L; $NaBH_4$, 1.0g/L)

The degradation of NBA increases with absorbance value and the reduction was around 60% whereas along with IGNC catalyst, the degradation was around 96% which indicates the kinetic barrier prevents the electron transfer from the donor BH_4^- to the acceptor NBA [99]. Then, the catalytic reduction of NBA was further exceptionally improved by the addition of $NaBH_4$ and the degradation of NBA in water had reached about 98%.

The absorbance peak was found to be shifted from 380 nm to 305 nm and 240 nm in the presence of $NaBH_4$, due to the formation of 4-phenyl-enediamine as shown in Fig. 5.9. The disappearance of the yellow color of the NBA solution was observed by UV–vis absorbance spectroscopy within 10 min after the addition of catalyst. It was clearly explained that the $NaBH_4$ acts as a hydrogen source which liberates hydrogen gas by hydrolysis at room temperature, and this process can be catalytically enhanced by metal ions [100]. Thus, the catalytic activity of the IGNC can be explained based on the morphology, particle size, since IGNC were significantly monodisperse and having a large surface area. This results in high efficiency of IGNC for the reduction of NBA. The catalytic reduction was found to be greatly dependent on solution pH, catalyst concentration and reducing agent concentration. Based on the results, it can be concluded that the sequential use of IGNC along with reducing agent $NaBH_4$ have remarkable advantage in the treatment of NBA containing wastewater.

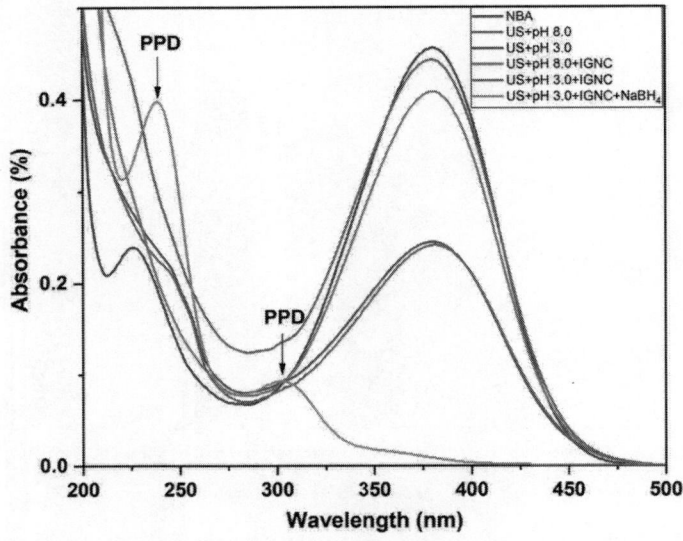

Fig. 5.9 UV–visible spectral variation of NBA at different experimental conditions

5.4   Synthesis of zerovalent iron–guar gum nanocomposite for the treatment of methyl orange

5.4.1   Synthesis procedure

For the synthesis of zerovalent iron oxide (ZVI)–guar gum nanocomposite (ZGNC), 1 M of $FeSO_4$ and AHG was subjected to ultrasound irradiation under nitrogen atmosphere. Then, 0.26 M of $NaBH_4$ aqueous solution was added drop wise at the rate of 3–15 mL/min after 15 min of ultrasonication. The reaction was continued and ultrasonication was halted when the solution turned from blue to black color which indicates the formation of the ZGNC. The as-synthesized nanocomposite was washed and dried.

5.4.2   Characterization of zerovalent iron nanocomposite

Thermo gravimetric analysis curve in Fig. 5.10 of AHG shows single step degradation. The weight loss at 59 °C might be due to loss of absorbed methanol. It starts to degrade at 100 °C. The polymer decomposition temperature has been found at 285 °C. The rate of weight loss increases with increase in temperature from 200 °C to 236 °C and thereafter decreases and attains a maximum value at about 350 °C. TGA analysis of zerovalent iron showed a maximum weight loss of 3% at 255 °C. The water loss is due to

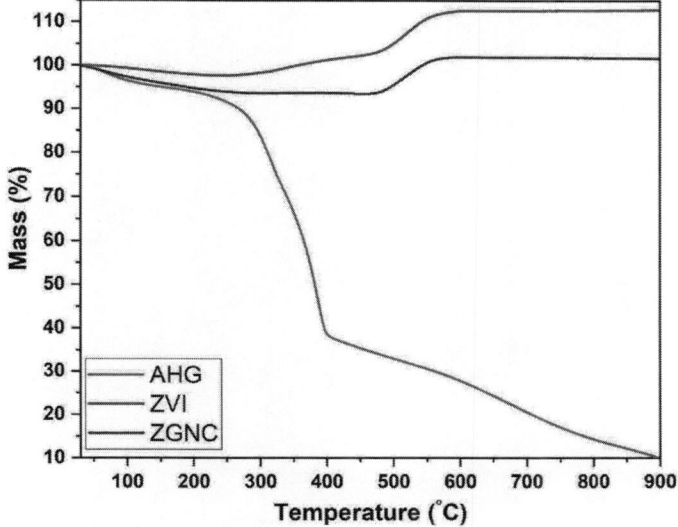

Figure 5.10   TGA thermograms of AHG, ZVI, and ZGNC

the surface contains 15% of the iron as Fe^{3+} [101] as shown in Fig. 5.10 which is persistent with the physical sorption of oxidized (surface exposed) Fe. The weight gain between 255 °C and 560 °C was due to the oxygen uptake of zerovalent iron and incompatible formation of hematite (α-Fe_2O_3). Figure 5.10 shows the thermograms of ZGNC, the weight loss at 255 °C was due to the presence of 15% of the iron as Fe^{3+}. The steady weight loss between 255 °C and 460 °C was due to the oxidative decomposition of the polymer, vaporization and elimination of volatile products. The thermograms of ZGNC reveals that the nanocomposite was more stable than the zerovalent iron nanoparticles.

FTIR spectra of AHG, ZVI, and ZGNC were shown in Fig. 5.11. FTIR spectra of GG in Fig. 5.11 illustrates the spectral regions between 2800 cm^{-1} and 3250 cm^{-1} are C–H stretching modes [91,92]. The absorption band due to ring stretching of mannose may be appeared at 1650–1700 cm^{-1}. The absorption bands recorded at 1382 cm^{-1} and 1358 cm^{-1} are may be due to symmetrical deformations of –CH$_2$ and –COH groups. Figure 5.11 shows the characteristic asymmetrical band region at 500–550 cm^{-1} indicates the Fe–O and Fe–OH stretching vibrations [95]. Figure 5.11 describes the peak observed in the region around 3400–3600 cm^{-1} was due to O–H stretching vibration of polymer and water involved in hydrogen bonding [93] which indicates the presence of guar gum present in the ZGNC. The absorption

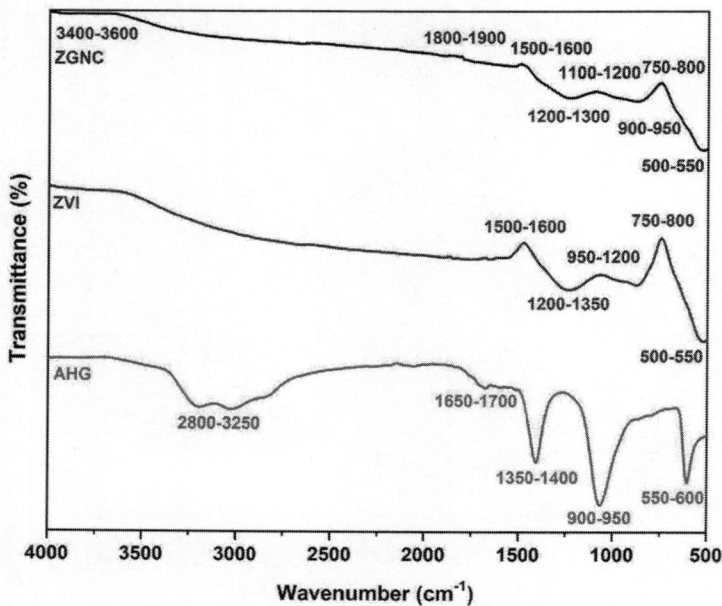

Figure 5.11   FTIR spectra of AHG, ZVI, and ZGNC

band due to ring stretching of mannose may be appeared at 1600 cm⁻¹. The absorption bands recorded at 1382 cm⁻¹ and 1358 cm⁻¹ are due to symmetrical deformations of $-CH_2$ and $-COH$ groups. The peaks observed in the spectra between 900 cm⁻¹ and 1150 cm⁻¹ represented the highly coupled C–C–O, C–OH, and C–O–C stretching modes of polymer backbone [92]. The FTIR spectral bands reveal the functional groups of the AHG were observed in ZGNC, which confirms that the polysaccharide was present in the nanocomposites.

The X-ray diffraction pattern of as-synthesized ZVI nanoparticle and ZGNC sample are shown in Fig. 5.12. The broad peak reveals the existence of an amorphous phase of iron. Apparent peaks at the 2θ of 44.9° and 35.8° indicate the presence of both zero-valent iron (α-Fe) and iron oxide (FeO) crystalline phases, whereas the ZVI nanoparticle shows very broad XRD peaks due to the short range order structure (that is, amorphous nature) of the iron nano-phases, including the Fe(0)phase [102–104]. ZVI samples often provide very broad XRD peaks due to the short-range order structure (that is, amorphous nature) of the iron nano-phases, including the Fe(0) phase. However, ultrasound assisted synthesis can provide more crystallinity to Fe (0) [105].

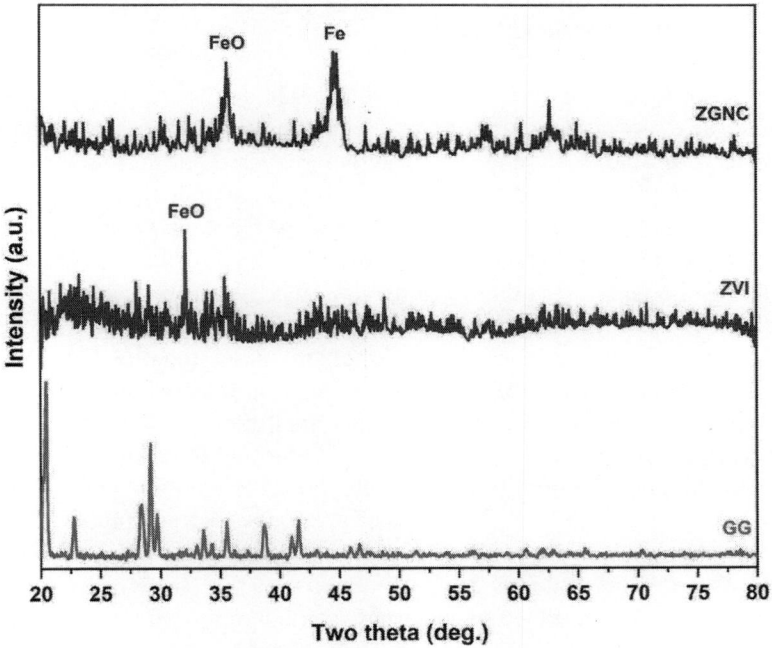

Figure 5.12   XRD patterns of AHG, ZVI, and ZGNC

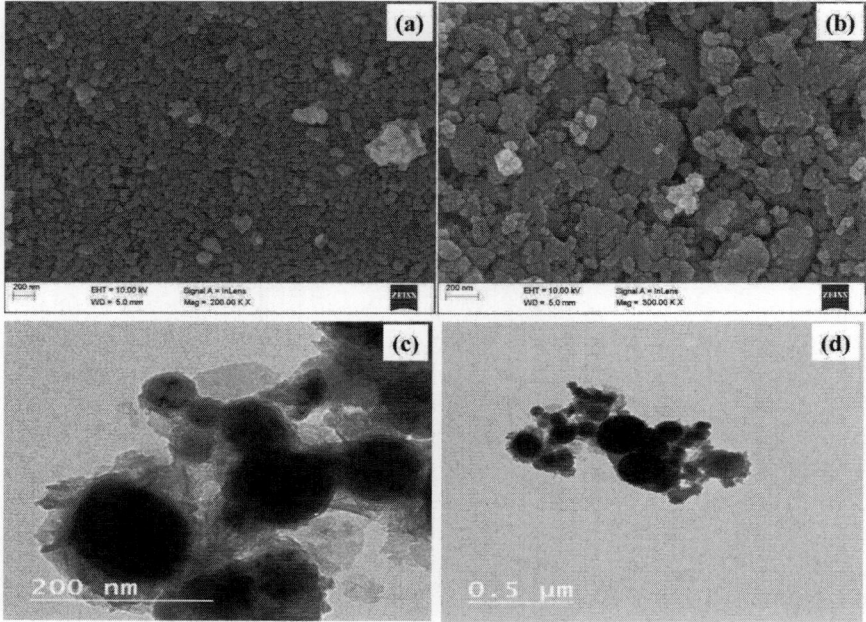

Figure 5.13 (a) SEM image of ZVI nanoparticle, (b) SEM image of ZGNC nanocomposite, (c and d) TEM images of ZGNC nanocomposite

SEM images from Fig. 5.13a and b reveals that these particles have a relatively narrow size distribution and nearly uniform spherical shape with individual particle size dimension less than 60 nm. The figure shows that the ZGNC improves the dispersion of nanoparticle and the aggregation of particle was reduced because of the presence of GG that weakens the magnetic interaction between particles. Figure 5.13c and d shows the low magnification TEM images of the as-prepared ZGNC, where the spherical shape with well crystallinity is revealed and the particles are connected in chains due to magnetic dipole interactions and chemical aggregation. The images show that ZVI are encapsulated by the GG polymeric networks (ZVI–AHG) structure. The formed AHG polymer networks expedite the excellent stability and the formation of the nanoparticle through electrostatic and steric effects and the hydroxyl groups of the polymeric chains, which further improve the stabilization of ZGNC [80,81; Pandey and Shivani, 2014]. The polymer GG act as a capping agent and hydroxyl groups of the polymeric chains and networks, they surround and protect the particles over longer periods. On the other hand, a strong physical adsorption of the AHG polymer onto the surface of the ZVI nanoparticle provides better stabilization of the particles formed.

5.4.3   Treatment of methyl orange by ZVI–guar gum nanocomposite

The catalytic activity of the ZGNC was studied with ZGNC and methyl orange (MO). The experiment was performed under magnetic stirring. The obtained decolorization efficiency was 99% as shown in Fig. 5.14. The removal efficiency was remarkable, and the solution was near to color-less. In order to contemplate that adsorption was not the main removal mechanism of MO, a 'OH radical scavenger (TBA) was introduced in the experiment. Figure 5.15 shows the effect of TBA on the removal of MO and the removal was inhibited by the presence of TBA. During oxidation, the Fenton reaction occurs in the presence of oxygen which likely forms a reactive oxygen species either on the particle surface or in the solution and strong oxidants are generated, such as hydroxyl radicals in the zerova-lent iron –H_2O system [106–108] as shown in Eqs. (5.1–5.3), which inten-sify the oxidation or degradation of organic pollutants in aqueous solution [109,110]. The reactions (5.1)–(5.3) might occur and the 'OH radical was generated indicating the oxidation reaction, which demonstrates that the 'OH radicals were responsible for MO decolorization. Thus, it can be concluded that oxidation would be the main removal mechanism for MO treatment.

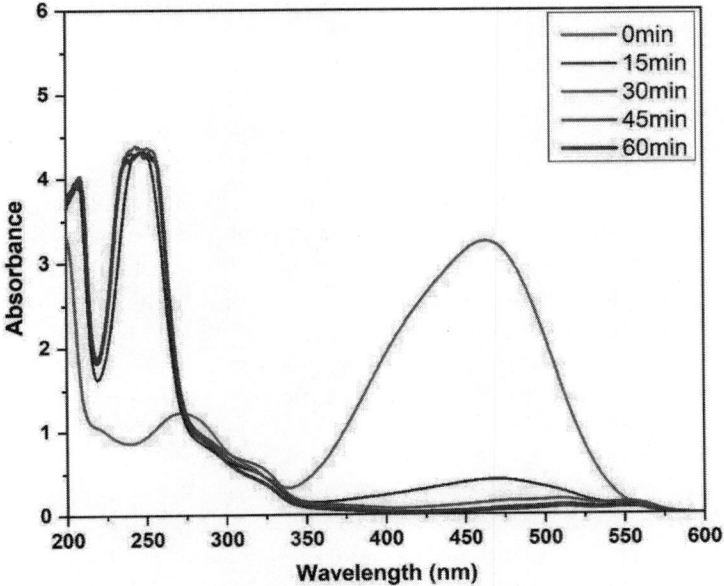

Figure 5.14   Decolorization of MO (pH=7.0±0.2, ZGNC=0.5 g/L, [MO]$_0$=100 ppm)

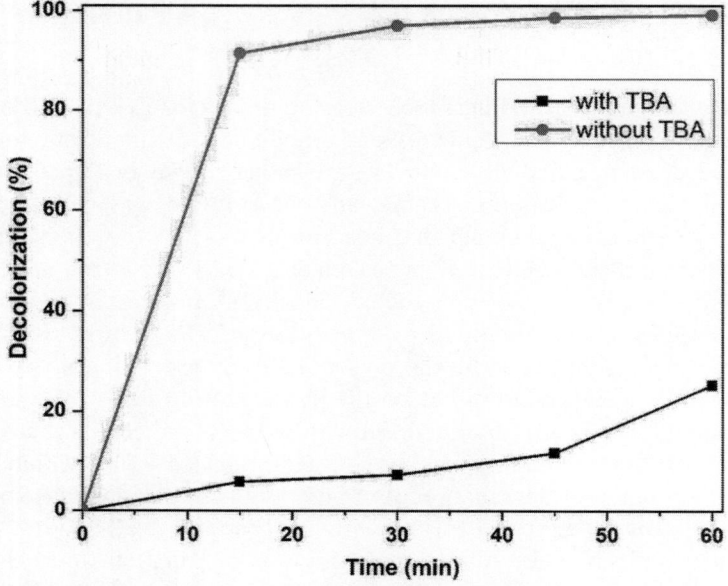

Figure 5.15 Effect of ·OH radical scavenger in the decolorization of MO (pH=7.0±0.2, ZGNC=0.5 g/L, [MO]$_0$=100 ppm, TBA=10 ppm)

$$Fe^0 + O_2 + 2H^+ \rightarrow H_2O_2 + 2Fe^{2+} \qquad (5.1)$$

$$Fe^{2+} + H_2O_2 \rightarrow Fe^{3+} + {}^{\cdot}OH + OH^- \qquad (5.2)$$

$$2Fe^{3+} + Fe^0 \rightarrow 3Fe^{2+} \qquad (5.3)$$

The as-synthesizes spherical ZGNC exhibits the dual characteristics of FeO and of zerovalent iron. The ZGNC was used for the catalytic oxidation of MO which resulted in removal efficiency of 99%. An enhanced performance of MO degradation was mainly due to the catalytic oxidation processes. A feasible mechanism for the degradation of MO by ZGNC was the oxidation of MO by hydroxyl radicals which was again confirmed by TBA. This study concludes that MO was significantly removed by Fenton-like process.

5.5   Conclusions

This chapter has focused on the recent advances in the preparation, characterization, and application of polsaccharide encapsulated metal and metal oxide nanocopmposites. Natural polysaccharides exhibited enormous potential in the green chemistry synthesis of metal and metal oxide nanocomposites owing to their nontoxic, biocompatible, and bio-degradable advantages. Polysaccharide can act as the reducing and sta-bilizing agents in the synthesis of metal and metal oxide nanoparticles, providing an alternative way of solving the problems conventional physi-cal and chemical synthesis approaches. The present chapter reported the sonochemical synthesis of iron oxide–guar gum and zerovalent iron–guar gum nanocomposites as a green chemistry approach with absence of non-toxic chemicals. The synthesized nanocomposites has a diverse nature of the functional group of the polysaccharide which controls the property of the hybrid nanomaterial. Different characterization techniques were performed to identify the presence of guar gum functional groups in the hybrid nanocomposites and the guar gum was encapsulated on the surface of the nanoparticle. The FESEM and HRTEM reveals the morphology of the nanocmposite, that is, as synthesized Fe_3O_4–guar gum nanocom-posite are cubic in shape with a partcle of 48 nm whereas the Fe^0–guar gum nanocomposite forms a chain like spherical shape with a particle size of ~60–70 nm. The synthesized IGNC have shown dual character-istics catalytic activity of oxidation and reduction for the treatment of p-nitroaniline.The catalytic reduction of NBA shows exceptional reduc-tion of 98% along with $NaBH_4$ which can undergo further treatment for complete oxidation. The as-synthesized ZGNC shows immense reponse for the catalytic oxidation of methyl orange which resulted in the removal efficiency of 99%. The reaction mechanism was determined by the addi-tion of ter-butyl alcohol (TBA), an •OH radical scavenger which displays strong oxidant radical was not generated and concludes the catalytic oxi-dation was the removal mechanism of MO. Therefore, based on the com-bination of the chemical nature, physical properties, and morphology of both the the functional polysacharide groups and the inorganic nanopar-ticles, it is possible to synthesis numerous hybrid metal and metal oxide nanocomposites with varied properties for further application of environ-mental remediation.

Abbreviations

The following abbreviations are used in this manuscript: Advanced oxida-tion process, AOPs; Guar gum, GG; Acid hydrolyzed guar gum, AHG; Iron oxide-guar gum nanocomposite, IGNC; Thermo gravimetric analysis, TGA; Fourier-transform infrared spectroscopy, FTIR; X-ray diffraction, XRD; Field emission scanning electron microscopy, FESEM; High resolution transmis-sion electron microscopy, HRTEM; 4-Nitrobenzeneamine, NBA; Methyl orange, MO; tert-Butyl alcohol, TBA.

References

1. Paul, S. A.; Chavan S. K.; Khambe, S. D. Studies on characterization of textile industrial wastewater in Solapur city. *Int. J. Chem. Sci.* **2012**, *10*, 635–642.

2. Gupta, V.; Suhas, K. Application of low cost adsorbents for dye removal: A review. *J. Environ. Manage.* **2009**, *90*, 2313–2342.

3. Mathur, N.; Bhatnagar, P.; Bakre, P. Assessing mutagenicity of textile dyes from Pali (Rajasthan) using AMES bioassay. *Appl. Ecol. Environ. Res.* **2005**, *4*, 111–118.

4. Gong, R.; Jin, Y.; Chen, J.; Hu, Y.; Sun, J. Removal of basic dyes from aqueous solution by sorption on phosphoric acid modified rice straw. *Dyes Pigments.* **2007**, *73*, 332–337.

5. Mane, V. S; Mall, I. D.; Srivastava, V. C. Use of bagasse fly ash as an adsorbent for the removal of brilliant green dye from aqueous solution. *Dyes Pigments.* **2007**, *73*, 269–278.

6. Hameed, B. H.; Ahmad, A. A.; Aziz, N. Isotherms, kinetics and thermodynamics of acid dye adsorption on activated palm ash. *Chem. Eng. J.* **2007**, *133*, 195–203.

7. Ali, H. Biodegradation of synthetic dyes: A review. *Water Air Soil Pollut.* **2010**, *213*, 251–273.

8. Zainith, S.; Sandhya, S.; Saxena, G.; Bharagava, R. N. Microbes an ecofriendly tool for the treatment of industrial waste waters. *Microbes Environ. Manage.* **2016**, 1, 78–103.

9. Hunger, K. In *Industrial Dyes: Chemistry, Properties, Applications*; **2003**. Cambridge, Wiley-VCH, Weinheim.

10. Bazin, I.; Hassine, A. I. H.; Hamouda, Y. H.; Mnif, W.; Bartegi, A.; Lopez-Ferber, M.; De Waard, M.; Gonzalez, C. Estrogenic and anti-estrogenic activity of 23 commercial textile dyes. *Ecotox. Environ. Saf.* **2012**, *85*, 131–136.

11. Aravind, U. K.; George, B.; Baburaj, M. S.; Thomas, S.; Thomas, A. P.; Aravindkumar, C. T. Treatment of industrial effluents using polyelectrolyte membranes. *Desalination.* **2010**, 252, 27–32.

12. Arivoli, S.; Thenkuzhali, M.; Prasath, M. D. Adsorption of rhodamine B by acid activated carbon: Kinetic, thermodynamic and equilibrium studies. *Orbital.* **2009**, *1*, 138–155.

13. Ju, D. J.; Byun, I. G.; Park, J. J.; Lee, C. H.; Ahn, G. H.; Park, T. J. Biosorption of a reactive dye (Rhodamine-B) from an aqueous solution using dried biomass of activated sludge. *Bioresour. Technol.* **2008**, *99*, 7971–7975.

14. Ding, S.; Li, Z.; Rui, W. Overview of dyeing wastewater treatment technology. *Water Resour. Prot.* **2010**, 26, 73–78.

15. Mani, S.; Bharagava, R. N. Isolation, screening and biochemical characterization of bacteria capable of crystal violet dye: Decolorization. *Int. J. Appl. Adv. Sci. Res.* **2017**, *2*, 70–75.

16. Turhan, K.; Durukan, I.; Ozturkcan, S. A.; Turgut, Z. Decolorization of textile basic dye in aqueous solution by ozone. *Dyes Pigments.* **2012**, *92*, 897–901.

17. Forgacs, E.; Cserhati, T.; Oros, G. Removal of synthetic dyes from wastewaters: A review. *Environ. Int.* **2004**, *30*, 953–971.

18. Manavi, N.; Kazemi, A. S.; Bonakdarpour, B. The development of aerobic granules from conventional activated sludge under anaerobic–aerobic cycles and their adaptation for treatment of dyeing wastewater. *Chem. Eng. J.* **2017**, *312*, 375–384.

19. Jadhav, A.J.; Srivastava, V. C. Adsorbed solution theory-based modeling of binary adsorption of nitrobenzene, aniline and phenol onto granulated activated carbon. *Chem. Eng. J.* **2013**, 229, 450–459.

20. Gupta, V. K.; Jain, R.; Agarwal, S.; Shrivastava, M. Kinetics of photo-catalytic degradation of hazardous dye Tropaeoline 000 using UV/TiO$_2$ in a UV reactor. *Colloid Surf. A: Physicochem. Eng. Aspects.* **2011a**, *378*, 22–26.

21. Gupta, V. K.; Jain, R.; Nayak, A.; Agarwal, S.; Shrivastava, M. Removal of the hazardous dye-tartrazine by photodegradation on titanium dioxide surface. *Mater. Sci. Eng. C.* **2011b**, *31*, 1062–1067.

22. Golob, V.; Ojstrsek, A. Removal of vat and disperse dyes from residual pad liquors. *Dyes Pigments.* **2005**, *64*, 57–61.

23. Mishra, A.; Bajpai, M. Flocculation behavior of model textile wastewater treated with a food grade polysaccharide. *J. Hazard. Mater.* **2005**, *118*, 213–217.

24. Yildiz, Y.; Koparal, A.; Keskinler, B. Effect of initial pH and supporting electrolyte on the treatment of water containing high concentration of humic substances by electrocoagulation. *Chem. Eng. J.* **2008**, *138*, 63–72.

25. Moreno, C. H. A.; Cocke, D. L.; Gomes, J. A.; Morkovsky, P.; Parga, J.; Peterson, E.; Garcia, C. Electrochemical reactions for electrocoagulation using iron electrodes. *Ind. Eng. Chem. Res.* **2009**, *48*, 2275–2282.

26. Tezcan, U. U.; Koparal, A. S.; Ogutveren, U. B. Electrocoagulation of vegetable oil refinery wastewater using aluminum electrodes. *J. Environ. Manage.* **2009**, *90*, 428–433.

27. Barrera-Diaz, C. E.; Lugo-Lugo, V.; Bilyeu, B. A review of chemical, electrochemical and biological methods for aqueous Cr(VI) reduction. *J. Hazard. Mater.* **2012**, *223*, 1–12.

28. Wu, C.; Kuo, C.; Chang, C. Decolorization of C.I. Reactive Red 2 by catalytic ozonation processes. *J. Hazard. Mater.* **2008**, *153*, 1052–1058.

29. Yeh, R. Y. L.; Hung, Y. T.; Liu, R. L. H.; Chiu, H. M.; Thomas, A. Textile wastewater treatment with activated sludge and powdered activated carbon. *Int. J. Environ. Stud.* **2002**, *59*, 607–622.

30. Alinsafi, A.; Khemis, M.; Pons, M. N.; Leclerc, J. P.; Yaacoubi, A.; Benhammou, A.; Nejmeddine, A. Electro-coagulation of reactive textile dyes and textile wastewater. *Chem. Eng. Process.* **2005**, *44*, 461–470.

31. Liu, C. H.; Wu, J. S.; Chiu, H. C.; Suen, S. Y.; Chu, K. H. Removal of anionic reactive 14 dyes from water using anion exchange membranes as adsorbers. *Water Res.* **2007**, *41*, 1491–1500.

32. Ciardelli, G.; Corsi, L.; Marcucci, M. Membrane separation for wastewater reuse in the textile industry. *Resour. Conserv. Recycl.* **2000**, *31*, 189–197.

33. Marcucci, M.; Nosenzo, G.; Capannelli, G.; Ciabatti, I.; Corrieri, D.; Ciardelli, G. Treatment and reuse of textile effluents based on new ultrafiltration and other membrane technologies. *Desalination.* **2001**, *138*, 75–82.

34. Crossley, C. How the dye industry is benefiting from membrane technology. *Filtr. Sep.* **2002**, *39*, 36–38.

35. Koyuncu, I.; Topacik, D.; Wiesner, M. R. Factors influencing flux decline during nano filtration of solutions containing dyes and salts. *Water Res.* **2004**, *38*, 432–40.

36. Tosik, R. Dyes color removal by ozone and hydrogen peroxide: Some aspects and problems. *Ozone Sci. Eng.* **2005**, *27*, 265–271.

37. Ledakowicz, S.; Maciejewska, R.; Gebicka, L.; Perkowski, J. Kinetics of the decolorization by Fenton's reagent. *Ozone Sci. Eng.* **2000**, 22, 195–205.

38. Hassan, D. U. B.; Aziz, A. R. A.; Daud, W. M. A. W. On the limitation of Fenton oxidation operational parameters: A review. *Int. J. Chem. React. Eng.* **2012**, *10*, 1542–6580.

39. Sheriff, T. S.; Cope, S. Ekwegh, M. Calmagite dye oxidation using in situ generated hydrogen peroxide catalysed by manganese(II) ions. *Dalton Trans.* **2007**, *5*, 119.

40. Parmon, V. Nanomaterials in catalysis. *Mater. Res. Innov.* **2008**, *12*, 60–66.

41. Liang, X. J.; Kumar, A.; Shi, D.; Cui, D. Nanostructures for medicine and pharmaceuticals. *J. Nanomater.* **2012**, 2.

42. Kusior, A.; Klich-Kafel, J.; Trenczek-Zajac, A.; Swierczek, K.; Radecka, M.; Zakrzewska, K. TiO_2–SnO nanomaterials for gas sensing and photocatalysis. *J. Eur. Ceram. Soc.* **2013**, *33*, 2285–2290.

43. Bujoli, B.; Roussiere, H.; Montavon, G. Novel phosphate–phosphonate hybrid nanomaterials applied to biology. *Prog. Solid-State Chem.* **2006**, *34*, 257–266.

44. Sarkar, S.; Guibal, E.; Quignard, F.; Sen, Gupta, A. K. Polymer-supported metals and metal oxide nanoparticles: Synthesis characterization and applications. *J. Nanopart. Res.* **2012**, *14*, 7–15.

45. Suslick, K. S. Sonochemistry. *Science.* **1990**, *247*, 1439–1445.

46. Suslick, K. S.; Doktycz, S. J. In *Advances in Sonochemistry*; **1990**. T J Mason, JAI Press, New York, *1*, 197.

47. Asghar, A.; Raman, A. A. A.; Daud, W. M. A. W. Advanced oxidation processes for in-situ production of hydrogen peroxide/hydroxyl radical for textile wastewater treatment: A review. *J. Clean Prod.* **2015**, *87*, 826–838.

48. Jadhav, A. J.; Holkar, C. R.; Karekar, S. E. Ultrasound assisted manufacturing of paraffin wax nanoemulsions: Process optimization. *Ultrason. Sonochem.* **2015**, *23*, 201–207.

49. Xu, H.; Zeiger, B. W.; Suslick, K. S. Sonochemical synthesis of nanomaterials. *Chem. Soc. Rev.* **2013**, *42*, 2555–26567.

50. Nishiyama, Y.; Langan, P.; Chanzy, H. Crystal structure and hydrogen-bonding system in cellulose Iβ from synchrotron X-ray and neutron fiber diffraction. *J. Am. Chem. Soc.* **2002**, 124, 9074–9082.

51. Scheller, H. V.; Jensen, J. K.; Sorensen, S. O.; Harholt, J.; Geshi, N. Biosynthesis of pectin. *Physiol. Plant.* **2007**, *129*, 283–295.

52. He, J. H.; Kunitake, T.; Nakao, A. Facile in situ synthesis of noble metal nanoparticles in porous cellulose fibers. *Chem. Mater.* **2003**, *15*, 4401–4406.

53. Chang, P. R.; Yu, J. G.; Ma, X. F. Fabrication and characterization of Sb_2O_3/carboxymethyl cellulose sodium and the properties of plasticized starch composite films. *Macromol. Mater. Eng.* **2009**, *294*, 762–767.

54. Yu, J. G.; Yang, J. W.; Liu, B. X.; Ma, X. F. Preparation and characterization of glycerol plasticized-pea starch/ZnO-carboxy-methylcellulose sodium nanocomposites. *Bioresour. Technol.* **2009**, *100*, 2832–2841.

55. Chang, P. R.; Yu, J.; Ma, X.; Anderson, D. P. Polysaccharides as stabilizers for the synthesis of magnetic nanoparticles. *Carbhyd. Polym.* **2011**, *83*, 640–644.

56. Lin, S. T.; Thirumavalavan, M.; Jiang, T. Y.; Lee, J. F. Synthesis of ZnO/Zn nano photocatalyst using modified polysaccharides for photodegradation of dyes. *Carbohydr. Polym.* **2014**, *105*, 1–9.

57. Hussain, M.; Shah, A.; Jantan, I.; Shah, M. R.; Tahir, M. N.; Ahmad, Bukhari, S. N. A hydroxy propyl cellulose as a novel green reservoir for the synthesis stabilization and storage of silver nanoparticles. *Int. J. Nanomed.* **2015**, 10, 2079–2088.

58. Chook, S. W.; Chia, C. H.; Chan, C. H.; Chin, S. X.; Zakaria, S.; Sajab, M. S.; Huang, N. M. A porous aerogel nanocomposite of silver nanoparticles-functionalized cellulose nanofibrils for SERS detection and catalytic degradation of rhodamine B. *RSC Adv.* **2015**, *5*, 88915–88920.

59. Ghasemzadeh, H.; Mahboubi, A.; Karimi, K.; Hassani, S. Full polysaccharide chitosan-CMC membrane and silver nanocomposite: Synthesis characterization and antibacterial behaviors. Polym. Adv. Technol. **2016**, 27, 1204–1210.

60. Zhao, C.; Li, J.; He, B.; Zhao, L. Fabrication of hydrophobic biocomposite by combining cellulosic fibers with polyhydroxy alkanoate. *Cellulose.* **2017**, *24*, 2265–2274.

61. Kokate, C. K.; Purohit, A. P.; Gokhale, S. B. Pharmacognosy; **2003**. 22nd ed, India, Nirali Prakashan, 133–166.

62. Villanueva, R. D.; Sousa, A. M. M.; Goncalves, M. P.; Nilsson, M.; Hilliou, L. Production and properties of agar from the invasive marine alga *Gracilaria vermiculophylla* (Gracilariales Rhodophyta). J. Appl. Phyco. **2010**, *22*, 211–220.

63. Shukla, M. K.; Singh, R. P.; Reddy, C. R.; Jha, B. Synthesis and characterization of agar-based silver nanoparticles and nanocomposite film with antibacterial applications. *Bioresour. Technol.* **2012**, 107, 295–300.

64. Makwana, D.; Castano, J.; Somani, R. S.; Bajaj, H. C. Characterization of Agar-CMC/Ag-MMT nanocomposite and evaluation of antibacterial and mechanical properties for packaging applications. *Arab. J. Chem.* **2018**. https://doiorg/ 101016/ jarabjc 2018.08.017.

65. Te-Wierik, G. H.; Eissens, AC.; Bergsma, J.; Arends-Scholte, A. W.; Bolhuis, G. K. A new generation starch product as excipient in pharmaceutical tablets III: Parameters affecting controlled drug release from tablets based on high surface area retrograded pregelatinized potato starch. *Int. J. Pharm.* **1997**, *157*, 181–187.

66. Budarin, V. L.; Clark, J. H.; Luque, R.; Macquarrie, D. J.; White, R. J. Palladium nanoparticles on polysaccharide-derived mesoporous materials and their catalytic performance in C–C coupling reactions. *Green Chem.* **2008**, *10*, 382–387.

67. Asharani, P. V.; Sethu, S.; Vadukumpully, S.; Zhong, S.; Lim, C. T.; Hande, M. P.; Valiyaveettil, S. Investigations on the structural damage in human erythrocytes exposed to silver gold and platinum nanoparticles. *Adv. Funct. Mater.* **2010**, *20*, 1233–1242.

68. White, R. J.; Budarin, V. L.; Moir, J. W. B.; Clark, J. H. A sweet killer: Mesoporous polysaccharide confined silver nanoparticles for antibacterial applications. *Int. J. Mol. Sci.* **2011**, *12*, 5782–5796.

69. Valodkar, M.; Rathore, P. S.; Jadeja, R. N.; Thounaojam, M.; Devkar, R. V.; Thakore, S. Cytotoxicityevaluation and antimicrobial studies of starch capped water soluble copper nanoparticles. *J. Hazard. Mater.* **2012**, *30*, 201–202:244–249.

70. Thirumavalavan, M.; Yang, F. M.; Lee, J. F. Investigation of preparation conditions and photocatalytic efficiency of nano ZnO using different polysaccharides. Environ. Sci. Pollut. Res. **2013**, 20, 5654–5664.

71. Chichova, M.; Shkodrova, M.; Vasileva, P.; Kirilova, K.; Doncheva-Stoimenova, D. Influence of silver nanoparticles on the activity of rat liver mitochondrial ATPase. *J. Nanopart. Res.* **2014**, *16*, 22–43.

72. Tripathy, T.; Kolya, H.; Jana, S.; Senapati, M. Green synthesis of Ag–Au bimetallic nanocomposites using a biodegradable synthetic graft copolymer; hydroxyethyl starch-g-poly(acrylamide-co-acrylic acid) and evaluation of their catalytic activities. *Eur. Polym. J.* **2017**, *87*, 113–123.

73. Barclay, T.; Ginic-Markovic, M.; Cooper, P.; Petrovsky, N. Inulin-a versatile polysaccharide with multiple pharmaceutical and food chemical uses. *J. Excip. Food Chem.* **2010**, *1*, 27–50.

74. Kalaivani, G. J.; Suja, S. K. TiO_2 (rutile) embedded inulin—A versatile bio-nanocomposite for photocatalytic degradation of methylene blue. *Carbohydr. Polym.* **2016**, *14*, 351–360.

75. Balachandramohan, J.; Anandan, S.; Sivasankar, T. A simple approach for the sonochemical synthesis of Fe_3O_4-guargum nanocomposite and its catalytic reduction of p-nitroaniline. *Ultrason. Sonochem.* **2018**, *40*, 1–10.

76. Doyle, J. P.; Lyons, G.; Morris, E. R. New proposals on hyperentanglement of galactomannans: Solution viscosity of fenugreek gum under neutral and alkaline conditions. *Food Hydrocol.* **2008**, *23*, 1501–1510.

77. Szopinski, D.; Luinstra, G. A viscoelastic properties of aqueous guar gum derivative solutions under large amplitude oscillatory shear (LAOS). *Carbohydr. Polym.* **2016**, *153*, 312–319.

78. Nandhini, K. V.; Abhilash, M. Study of hydration kinetics and rheological behaviour of guar gum. *Int. J. Pharm. Sci. Res.* **2010**, *1*, 28–39.

79. Coviello, T.; Alhaique, F.; Dorigo, A.; Matricardi, P.; Grassi, M. Two galactomannans and scleroglucan as matrices for drug delivery: Preparation and release studies. *Eur. J. Pharm. Biopharm.* **2007**, *66*, 200–209.

80. Pandey, S.; Goswami, G. K.; Nanda, K. K. Green synthesis of biopolymer-silver nanoparticle nanocomposite: An optical sensor for ammonia detection. *Int. J. Biol. Macromol.* **2012**, *51*, 583–589.

81. Pandey, S.; Goswami, G. K.; Nanda, K. K. Green synthesis of polysaccharide/gold nanoparticle nanocomposite: An efficient ammonia sensor. *Carbohydr. Polym.* **2013**, *94*, 229–234.

82. Pandey, S.; Nanda, K. K. Au nanocomposite based chemiresistive ammonia sensor for health monitoring. *ACS Sens.* **2016**, *1*, 55–62.

83. Rastogi, P. K.; Ganesan, V.; Krishnamoorthi, S. Palladium nanoparticles decorated gaur gum based hybrid material for electrocatalytic hydrazine determination. *Electrochim. Acta.* **2014**, *125* 593–600.

84. Das, T.; Yeasmin, S.; Khatua, S.; Acharya, K.; Bandyopadhyay, A. Influence of a blend of guar gum and poly(vinyl alcohol) on long term stability and antibacterial and antioxidant efficacies of silver nanoparticles. *RSC Adv.* **2015**, *5*, 54059–54069.

85. Ye, X.; Lin, D.; Jiao, Z.; Zhang, L. The thermal stability of nanocrystalline maghemite Fe_2O_3. *J. Phys. D: Appl. Phys.* **1998**, *31*, 2739–2744.

86. Lauer, Jr. H. V.; Ming, D. W.; Golden, D. C. Thermal analysis of acicular shaped magnetite. *Lunar Planetary Sci.* **2003**, XXXIV. http://wwwlpiusraedu/meetings/lpsc2003/pdf/1341pdf.

87. Maity, D.; Agrawal, D. C. Synthesis of iron oxide nanoparticles under oxidizing environment and their stabilization in aqueous and non-aqueous media. *J. Magn. Magn. Mater.* **2007**, 308, 46–55.

88. Sanders, J. P.; Gallagher, P. K. Kinetics of the oxidation of magnetite using simultaneous TG/DSC. *J. Therm. Anal. Calorim.* **2003**, *72*, 777–89.

89. Fajaroh, F.; Setyawan, H.; Nur, A.; Lenggoro, I. W. Thermal stability of silica-coated magnetite nanoparticles prepared by an electrochemical method. *Adv. Powder Technol.* **2013**, *24*, 507–511.

90. Lepp, H. Stages in the oxidation of magnetite. *Am Miner.* **1957**, *42*, 679–681.

91. Mudgil, D.; Barak, S.; Khatkar, B. S. X-ray diffraction IR spectroscopy and thermal characterization of partially hydrolyzed guar gum. *Int. J. Biol. Macromol.* **2012**, *50*, 1035–1039.

92. Kacurakova, M.; Belton, P. S.; Wilson, R. H.; Hirsch, J.; Ebringerova, A. Hydration properties of xylan-type structures: An FTIR study of xylo oligosaccharides. *J. Sci. Food Agric.* **1998**, *77*, 38–44.

93. Fringant, C.; Tvaroska, I.; Mazeau, K.; Rinaudo, M.; Desbrieres, J. Hydration of alpha-maltose and amylose: Molecular modelling and thermodynamics study. *Carbohydr. Res.* **1995**, *278*, 27–41.

94. Kacurakova, M.; Ebringerova, A.; Hirsch, J.; Hromadkova, Z. Infrared study of arabinoxylans. *J. Sci. Food Agric.* **1994**, *66*, 423–427.

95. Stoia, M.; Istratie, R.; Pacurariu, C. Investigation of magnetite nanoparticles stability in air by thermal analysis and FTIR spectroscopy. *J. Therm. Anal. Calorim.* **2016**, *125*, 1185–1198.

96. Hong, R. Y.; Feng, B.; Chen, L. L.; Liu, G. H.; Li, H. Z.; Zheng, Y.; Wei, D. G. Synthesis, characterization and MRI application of dextran-coated Fe_3O_4 magnetic nanoparticles. *Biochem. Eng. J.* **2008**, *42*, 290–300.

97. Silva, V. A. J.; Andrade, P. L.; Silva, M. P. C.; Bustamante, A. D.; De Los Santos Valladares, L.; Albino Aguia, J. Synthesis and characterization of Fe_3O_4 nanoparticles coated with fucan polysaccharides. *J. Magnet. Magnet. Mater.* **2013**, *343*, 138–143.

98. Sun, J. H.; Sun, S. P.; Fan, M. H. A kinetic study on the degradation of p-nitroaniline by Fenton oxidation process. *J. Hazard. Mater.* **2007**, *148*, 172–177.

99. Abbas, M.; Torati, S. R.; Kim, C. A novel approach for the synthesis of ultrathin silica-coated iron oxide nanocubes decorated with silver nanodots (Fe_3O_4/SiO_2/Ag) and their superior catalytic reduction of 4-nitroaniline. *Nanoscale.* **2015**, *7*, 12192–12204.

100. Rakap, M.; Ozkar, S. Hydroxyapatite-supported cobalt (0) nanoclusters as efficient and cost-effective catalyst for hydrogen generation from the hydrolysis of both sodium borohydride and ammonia–borane. *Catal. Today.* **2012**, *183*, 17–25.

101. Ponder, S. M.; Darab, J. G.; Bucher, J.; Caulder, D.; Craig, I.; Davis, L.; Edelstein, N.; Lukens, W.; Nitsche, H.; Rao, L.; Shuh, D. K.; Mallouk, T. E. Surface chemistry and electrochemistry of supported zerovalent iron nanoparticles in the remediation of aqueous metal contaminants. *Chem. Mater.* **2001**, *13*, 479–486.

102. Kanel, S. R.; Manning, B.; Charlet, L.; Choi, H. Removal of arsenic (III) from groundwater by nanoscale zero-valent iron. *Environ. Sci. Technol.* **2005**, *39*, 1291–1298.

103. Wang, Q.; Kanel, S.; Park, H.; Ryu, A.; Choi, H. Controllable synthesis characterization and magnetic properties of nanoscale zerovalent iron with specific high Brunauere Emmette Teller surface area. *J. Nanopart. Res.* **2009**, *11*, 749–755.

104. Fang, Z.; Chen, J.; Qiu, X.; Qiu, X.; Cheng, W.; Zhu, L. Effective removal of antibiotic metronidazole from water by nanoscale zero-valent iron particles. *Desalination.* **2011**, *268*, 60–67.

105. Jamei, M. R.; Khosravi, M. R.; Anvaripour, B. A novel ultrasound assisted method in synthesis of NZVI particles. *Ultrason. Sonochem.* **2014**, *21*, 226–233.

106. Chang, S.; Wang, K.; Chao, S.; Peng, T.; Huang, L. Degradation of azo and anthraquinone dyes by a low-cost FeO/air process. *J. Hazard. Mater.* **2009**, *166*, 1127–1133.

107. Noradoun, C. E.; Cheng, I. F. EDTA degradation induced by oxygen activation in a zerovalent iron/air/water system. *Environ. Sci. Technol.* **2005**, *39*, 7158–7163.

108. Joo, S. H.; Feitz, A. J.; Sedlak, D. L.; Waite, T. D. Quantification of the oxidizing capacity of nanoparticulate zero-valent iron. *Environ. Sci. Technol.* **2005**, *39*, 1263–1268.

109. Noradoun, C. E.; Engelmann, M. D.; McLaughlin, M.; Hutcheson, R.; Breen, K.; Paszczynski, A.; Cheng, I. F. Destruction of chlorinated phenols by dioxygen activation under aqueous room temperature and pressure conditions. *Chem. Res.* **2003**, *42*, 5024–5030.

110. Joo, S. H.; Feitz, A. J.; Waite, T. D. Oxidative degradation of the carbothioate herbicide molinate using nanoscale zero-valent iron. *Environ. Sci. Technol.* **2004**, *38*, 2242–2247.

6

Textile effluent treatment by electrochemical methods: A critical review

Shalini V and Susmita Das*

*Department of Chemical Engineering, National Institute of Technology Calicut,
Calicut 673601, India*
Corresponding Author. E-mail: susmita.82.das@gmail.com.
Tel.: +91-4952-285471.

Abstract: Electrochemical methods provides a clean alternative to the degradation and decolourization of chemically stable dyes in textile effluent water by using electron as active chemical agent. The recalcitrant pollutants could be efficiently treated by moderate amount of current without production of any solid residue. This article reviews the detail progress in electrochemical method in textile effluent treatment specifically dye degradation. Importantly, conventional methods are inefficient to degrade reactive dye such as azo dye which could resolve by carrying out degradation with electrochemical method. This paper discusses several potential electrochemical techniques for dye degradation.

Keywords: Dye degradation; electrochemical methods; textile effluent treatment.

6.1   Introduction

Textile industry, an important part of our civilization, produces large quantity of waste water during the dyeing and finishing processes. It is a real challenge for researches to treat the textile effluent water efficiently [1]. Among all other pollutant within effluent, synthetis dye has gained a lot of attention [2]. Most widely used reactive dyes are azo group molecules which have complex aromatic compounds with large structural diversity which yield very high degree of chemical, biological, and photocatalytic stability and breakdown resistance with exposure to sunlight, microorganisms, water and soap, etc [3,4]. As per literature, the conventional methods for dye degradation includes physiochemical (coagulation/flocculation, adsorption, and membrane separation) [2,5,6], chemical (ozonation and oxidation with hypochlorite ion) [1,2,5,7] and biological (anaerobic process, oxidation, trickling filters and activated sludge process) [7,8] process. These methods are cheap and simple for dye removal but inefficient to degrade azo dye and also it producess colourless toxic aromatic amine compounds as a byproduct [7]. These problems have been addressed by reaserchers and solved by using aspecific treatment called electrochemical treatment. As compare to convenvention methods, electrochemical methods are more economical, easy to handle, versatile and clean alternative for textile effluent treatment [8]. Electrochemical method has also provided high dye degradation efficiency, it can be operated at moderate temperature and furthermore does not require chemical reagent and also prevent secondary byproduct [6]. Before 2000, very less number of researches has reported dye degradation by electrochemical treatment method but during last few year researchers have shown a great interest in this methods. Electrochemical treatment techniques such as electrochemical oxidation [9], electrochemical reduction [10], electrocoagulation [11] and electrocoagulation/flotation [12], electrodialysis [13], and electrochemical advanced oxidation processes [14], etc. has been widely used for dye degradation specifically, degradation of non-biodegradable dyes such as reactive dyes [15–17].

The present article reviews the progress of scientific research on electrochemical method for textile effluent water treatment specifically for batch scale dye degradation by compiling data on various electrochemical methods, electrodes used, dye degraded and factors influencing the degradation process.

6.2   Electrochemical methods for textile effluent industry

Until now, various types of electrochemical techniques such as electrochemical coagulation, electrochemical oxidation, electrochemical reduction, and photoelectrocatalysis have been used for treatment of textile effluent water specifically dye degradation. In the next section, few important and most used electrochemical techniques have been discussed.

6.2.1   Electrocoagulation

In general, in electrocoagulation, metal cations have been formed by electrodissolution of the anodes. Simultaneously, water is reduced to hydrogen gas and the hydroxyl ion. The cations destabilize colloidal particles by neutralizing charges. The monomeric and polymeric hydroxide complex species generated, progressively degrades the dyes [18–20]. Several other electrocoagulation method such as active chlorine method, sono-electrocoagulation, peroxi-coagulation and photo-electrocoagulation method has also been studied, where active radicals has been generated with the help of UV source or hydrogen peroxide or chlorine, etc. In the following section, we will focus on recent achievements dealing with the development of effective strategies for degradation of dye in all kind of electrocoagulation method.

Bassyouni et al. [21] degraded ~87% di-azo dye, acid brown 14 by using batch mode electrocoagulation method. The COD removal was ~48.7%. Iron sheet lining was prepared on the inner wall and bottom part of an open and undivided plexi-glass cylindrical cell of 2.5 L capacity which was used as the anode. The iron sheet cathode of similar shape was placed at a distance of 1 cm from the anode. Effects of various experimental parameters, for example, initial dye concentration, current density, pH, electrolyte concentration, on dye degradation has been studied. Assamian et al. [22] degraded methylene blue by electrocoagulation process as shown in Fig. 6.1. Iron and aluminum electrodes were used for experimental purpose. Result showed that at optimum condition 89% methylene blue has been degraded by using a factorial experimental design method. Ghernaout et al. [23] performed electrocoagulation experiment for degrading direct brown 2 and BF cibacete blue dye. Aluminum sheets were used as the electrodes which were kept within the electrolyte solution. A degradation efficiency of 99.9% and 99.7% has been achieved for direct brown 2 and BF cibacete blue dye, respectively, for batch mode operation. Effects of pH, supporting electrolyte concentration (NaCl),

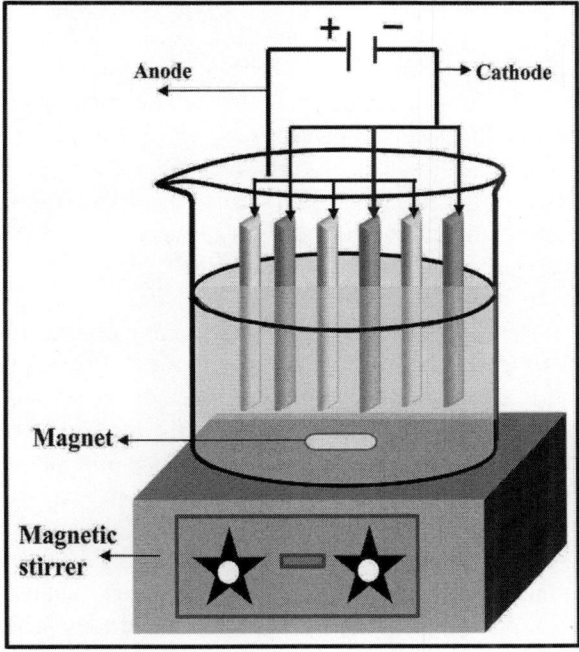

Figure 6.1 Experimental set-up for electrocoagulation method [22]

inter electrode distance, and power consumption for optimal degradation efficiency has been studied. Gilpavas et al. [24] reported the degradation of indigo dye by combination of electrocoagulation and electrooxidation method giving a combined color removal efficiency of ~100% and COD decrease of 71.4% with the help of a fractional factorial design of experimental method. Current density, pH, and conductivity of the solution were the parameters under consideration. Fe or Al sacrificial anodes were coupled with titanium cathode in a Plexiglas reactor. Ardhan et al. [25] degraded ~98% reactive blue 21 by electrocoagulation method. The anode assembly consisted of iron scraps from workshop turnings in PVC housing. The metal ion transport was facilitated with the provision of 5 mm holes made on the housing. The cathodes were made of stainless steel. The effect of current variation and electrocoagulation time on dye removal efficiency has been discussed and result also showed that the COD removal efficiency was 93%. Lopez et al. [26] designed cartridge type electrocoagulation reactor using steel wool anode and aluminum rod cathode for the degradation of ramazol red RB 133. A dye removal efficiency of ~99% was reported depending on the current density.

Fu [27] reported that, the major challenge during electrocoagulation is electrode surface passivation, polarization, and dissolution of the sacrificial anode resulting in subsequent increase in the metal concentration in treated effluent. An asynchronous current supply switching between iron and aluminum electrodes was suggested to mitigate these effects. This has been resolved by using metallic iron and aluminum electrodes in organic glass electrolyser. The passivation film formed on the anode has been removed when the anode becomes cathode in the reversed current flow cycle. Result showed a higher degradation efficiency ~96% for color removal and 96% for COD for reactive brilliant blue X-BR dye. Duan et al. [28] studied the discoloration of synthetic waste water containing methylene blue using periodic electrode reversal method with response surface methodology (RSM). At optimum conditions, a removal efficiency of ~97% has been observed. Effect of various experimental parameters like cell voltage, current density, and electrode reversal period has been studied. Hendaoui et al. [29] degraded ~92.9% real indigo dye by electrocoagulation method. A two chamber, bottom fed, continuous reactor made of high density poly ethylene was equipped with iron electrodes placed at a distance of 4 mm. The effects of three parameters like initial pH, applied voltage between the electrodes, and inlet flow rate have been studied.

Hosiny et al. [30] evaluated the performance of electrocoagulation process combined with flotation for dye degradation. Experiment was carried out by using aluminum electrodes within a cylindrical glass shell reactor where the distance of 4 cm was kept in-between the electrodes. The degradation dependence on iconicity of acid red 1 (anionic), basic violet 3 (cationic) and disperse blue 14 (non-ionic) were monitored in a cylindrical glass shell reactor using two aluminum electrodes kept at an inter electrode distance of 4 cm. 0.2 % KCL was added to the solution to prevent electrode passivation. The order of dye removal efficiency was ~91 %, 78 % and 59 % for non-ionic, cationic and anionic dyes, respectively. The COD removal efficiency of the real textile waste water containing the three dyes was found to be 92%. Acid red 336 was degraded with a removal efficiency of 95 % using a continuous photovoltaic electrocoagulation by Khemila et al. [31]. Aluminum electrodes spaced at a distance of 1 cm in the electrolyser. The photovoltaic cells were considered to be of single diode model. Apart from the process parameters inherent to the continuous electrocoagulation process, the effects of solar irradiation and energy consumption were reported.

Naje et al. [32] designed a solar power driven electrocoagulation with rotating anode for the removal of reactive blue 19 .The removal efficiencies for the batch and continuous process were 95 % and 95.5 %, respectively. Aluminum electrodes have been used for the experimental purpose. The rotating anode was composed of impellers with cathode rings encircling the anode. The effects of rotational speed of the anode, current density, recirculation flow rate, residence time, and continuous flow regime on the removal efficiency has been studied. Silva et al. [33] reported the effects of solar irradiation, current density and flow rate on the degradation of synthetic textile waste water containing an azo dye, remazol yellow. The reactor made of two aluminum electrodes formed by 39 circular blades spaced 1 cm from each other. The removal efficiency was found to be 70 % for color and 84.8 % for COD in a continuous acrylic reactor of 17 L capacity. Phalakornkule et al. [34] performed a continuous-flow sparged packed-bed electrocoagulator for reactive blue 21 degradation. A perforated PVC pipe stacked with iron nuts were used as anode. The assembly included three packed bed anodes and four tubular cathodes made of stainless steel. The hydrogen generated during the oxidation reduction reaction was recirculated and sparged in to the center of the anode through sparging tubes. The advantages of hydrogen recirculation included higher color removal efficiency, reduced cell resistance and decreased energy consumption for homogenization. The COD removal was ~99 % and color degradation efficiency was ~98.4 % and ~99.8 % for with and without gas sparging condition, respectively. Assemian et al. [35] degraded bio-recalcitrant synthetic dye in a batch mode operation by electro-coagulation method with the help of factorial experimental design. Iron and aluminum electrodes were used for experimental purpose. Effect of various experimental parameters such as dye concentration, electrolysis time, current intensity and electrode type on dye degradation has been studied. Result illustrated that at optimum condition ~89 % methylene blue has been degraded. In another report, author has carried out degradation of textile bio-refractory wastewater by similar process where iron electrodes have been used for the experimental purpose. Experiments have been optimized by one-factor-at-a time methodology and result illustrated that ~84 % dye has been degraded.

Electroflotation is a process similar to electrocoagulation. Most often electrocoagulation and electroflotation occurs simultaneously in the same cell. For example, Zodi et al. [12] carried out degradation of direct red 81 dye within a continuous flow electrocoagulation/flotation reactor. A Pyrex reactor with aluminum electrodes has been designed for experimental purpose. Result showed a degradation efficiency of ~85.4%.

6.2.2   Electrochemical oxidation method

Electrochemical oxidation of dyes follows either of the following two methods: direct oxidation or mediated oxidation. During direct oxidation, the pollutants are directly degraded at the anode with the help of power supply and electrodes. The process is limited to the surface of the anode and is also known as anodic oxidation [36,37]. In mediated oxidation, an external active chemical species has been supplied to the system. In the next section, few important dye degradation by electrochemical oxidation process has been discussed.

6.2.2.1   Direct oxidation method

Electrochemical direct oxidation proceeds in two different ways. The oxidation can be resulted by the direct transfer of electrons from the anode to the dye molecules or through the reactive species formed on the surface of the anode [38]. Abdessamad et al. [39] carried out anodic oxidation of real textile effluents containing a mixture of dyes using boron doped diamond (BDD) electrodes. Monopolar and bipolar cell has been used for a comparative analysis of effluents. The monopolar cell consisted of a monopolar-p-silicon covered with BDD as anode and a stainless steel as cathode. In the bipolar cell, two bipolar Si/BDD coated electrodes was placed between two monopolar electrodes. The decoloration efficiency was 74 % and 65 % for monopolar and bipolar cell, respectively and the corresponding COD removal efficiency was 68 % and 100 %. Hamad et al. [40] degraded reactive violet 2 and acid brown 14 dye by using graphite anode and cylindrical stainless steel cathode. At optimum operating conditions the degradation efficiency of reactive violet 2 and acid brown 14 dye were 95 % and 69.8 %, respectively. Soares et al. [41] degraded textile azo dyes (reactive orange 16, reactive violet 4, reactive red 228, and reactive black) by using BDD anode and carbon polytetrafluoroethylene air diffusion electrode as cathode. The degradation efficiencies were 95.5 %, 98.5 %, 74.6 %, and 85.5 % for reactive orange 16, reactive violet 4, reactive red 228, and reactive black dye, respectively. Sanchez et al. [42] has also used BDD electrode for treating real textile effluent. Five BDD electrodes were placed vertically inside a batch reactor. Two of them were utilized as cathodes and other three as anodes. Under optimum conditions, COD reduction and decoloration efficiency was 41.9 % and 60.8 %, respectively. Rao et al. [43] conducted electrochemical oxidation of real textile waste water using a three electrode system. Ti/TiO$_2$-NTs/Sb-SnO$_2$ anode synthesized by electrodeposition was used along with Pt auxiliary electrode and Ag reference electrode has been used for experimental purpose. Result showed COD removal efficiency of ~41 % and dye degradation efficiency ~52.83 %. Xu et al. [44] degraded acid red 73 by using direct oxidation method where Nb-Ti/Nb-TiO$_2$-NTs/SnO$_2$-Sb used as anode and Ti as cathode. The color and COD removal efficiency was reported as 85 % and 67.7 %, respectively. The degradation of Acid 28 using Ti/b-PbO$_2$ and Ti-Pt/b-PbO$_2$ anodes and Ni plate cathode was reported by Irikura et al. [45] giving 90 % discoloration, ~100 % COD removal and 30–33 % TOC removal after 3 h of electrolysis. GilPavas et al. [46] degraded acid yellow 23 dye by anodic oxidation method. Monopolar electrodes of graphite anode and titanium cathode were used for experiments. At the optimum operating conditions predicted by RSM (Response surface methodology) coupled with Box-Behnken experimental design (BBD), removal efficiencies of color and COD was ~99 % and ~76 %, respectively. Real textile effluent degradation using Ti/RuO$_2$ anode and aluminum plate cathode reported by Kaur et al. [47]. Box behnken design under RSM was use to optimize the process parameters such as current, pH, and electrolysis time. Under optimum condition COD and color removal efficiency was 80.0 % and 97.3 %, respectively. Ferreira et al. [48] reported the degradation of novacron blue using a three electrode arrangement in a dual flow cell. The working electrode was Ti/Pt-Ti/Pt-SbSn, platinum wire was utilized as auxiliary electrode and Ag/AgCl/KCl as reference electrode. An optimum operating condition, the COD and color removal efficiencies were reported as ~60 % and ~100 %, respectively.

6.2.2.2   Mediated oxidation method

During mediated electrochemical oxidation, an external reagent is added to the electrolytic cell. The dissociation of the reagent produces strong oxidizing agents, which successively degrades the dye molecule [49]. The most widely used mediated oxidation methods are active chlorine method and electro Fenton method [7]. Rodriguez et al. [50] reported the degradation of synthetic dye containing indigo carmine by active chlorine method using Sb$_2$O$_5$ doped Ti/RuO$_2$-ZrO$_2$ Electrode with electro generation of active chlorine in a filter press type reactor. Rajkumar and Kim [51] degraded various dye including refraction red, refazol red, reactive black, reactive yellow, and reactive orange by mediated oxidation method. Titanium mesh coated with oxides of titanium, ruthenium and iridium used as anode electrode and stainless steel as cathode. The result exhibited ~100 % dye degradation efficiency and ~73.5 % COD reduction capability. Further, Rajkumar and Muthukumar [52] again carried out degradation of reactive orange 107 dye by using graphite carbon electrode as anode and cathode. Experimental result showed a decoloration efficiency of 98 % and a COD reduction of ~90 %.

Fenton chemistry based textile effluent treatment such as electro Fenton process (EF), UV Photoelectro Fenton process (PEF), and solar photoelectron Fenton (SPFE) process has been one of the most promising electrochemical advanced oxidation processes [53]. Antonin et al. [54] carried out a comparative study of EF and SPEF method on degradation of di-azo dye like Evans blue. EF process performed within a stirred tank reactor with boron doped diamond electrode as anode and porous graphite air diffusion cathode showed ~100 % color and 89 % COD removal efficiency. Whereas, SPEF process shows color and TOC removal efficiency was 98 % and 88 % respectively.

Ghosh et al. [55] performed degradation of methylene blue and titan yellow dye by using photoelectro Fenton method where graphite and iron electrode were used as cathode and anode. Result showed 98 % and 80 % discoloration efficiency for methylene blue and titan yellow dye, respectively with NaCl as support electrolyte. Gilpavas et al. [56] designed a solar driven photoelectro Fenton process for treatment of real textile waste water by using RMS optimization. Experiments were carried out in a batch reactor by using Si/boron doped diamond electrode as anode and titanium electrode as cathode. At optimum conditions ~100 % discoloration, 83 % COD reduction, and 70 % TOC reduction efficiency has been shown. The effects of pH, current density, conductivity, Fe^{2+} concentration, and anode area on the dye degradation efficiency have been studied. Iglesias et al. [57] carried out degradation of reactive black 5 by electro-Fenton process using Fe alignated gel beads. Two graphite electrode sheets spaced at 6 cm was placed in Fe^{2+} contain dye solution. At optimum predicted by statistical analysis using ANOVA decoloration and COD reduction removal efficiency were 90 % and 65.3 %, respectively. Further, Kaur et al. [47] also treated textile wastewater by electro Fenton process. Two Ti/RuO_2 electrodes and two aluminum electrodes were used as anode and cathode within a cubical plexi glass reactor. Malakootian and Moridi [58] reported the removal efficiency of synthetic and real azo dye (acid dye 18) by using electro Fenton process where two iron electrodes were used for experimental purpose. Result showed that at optimum condition, the discoloration efficiency has found ~100 % and 90.5 % for synthetic dye and real textile effluent, respectively. The effect of pH, voltage, hydrogen peroxide concentration, initial dye concentration, type and concentration of electrolyte, distance between electrodes, and electrolysis time were examined. Pajootan et al. [59] investigated the usability of carbon nanotubes electrodes coated with immobilized TiO_2 for dye degradation in a continuous photocatalytic electro Fenton process in a PEF-TiO_2 bubble reactor. Degradation of binary solution containing acid red 14 and acid blue 92 was carried out using graphite electrodes as cathode and anode under UV irradiation. At optimum condition, the discoloration of acid red 14 and acid blue 92 has been observed ~65.67 % and 74.5 %, respectively. The binary solution showed a COD reduction of 63.7 %. Whereas, NaCl assisted process showed decoloration efficiency ~68.8 % and ~80.3 % for acid red 14 and acid blue 92 dye, respectively. Shwe et al. [60] showed the influence of metal oxide coated electrodes on the decoloration efficiency of textile waste water containing reactive blue 19 (Remazol brilliant blue) by electro Fenton process. RuO_2 and IrO_2 mixed electrodes were used for experimental purpose. For NaCl assisted process, at optimum conditions result showed ~100 % decoloration efficiency of the dye. Urmi et al. [61] has also observed ~90 % methyl orange dye degradation by using photoelectro Fenton method.

6.2.3   Electrocatalysis and photoelectrocatalysis

Electrocatalytic method utilizes the catalytic properties of specific reagents to accelerate the oxidation of organic pollutants in waste water. Conventional electrocatalytic methods uses anode modification with transition metals using various nano synthesis techniques. Zhang et al. [62] degraded alizarin yellow R dye by using Ti/PbO_2–Sm_2O_3 composite based electrodes. Anode fabricated by simple electro deposition method was used in conjunction with stainless steel cathode. Several experimental parameters like electrolytic concentration, plate spacing, initial pH, and cell voltage has been studied and at an optimum condition, result showed a maximum degradation efficiency ~80%. He et al. [63] degraded alizarin yellow R wastewater by using electrocatalytic method. Ti/PbO_2–Sm_2O_3 composite electrode prepared via electrodeposition method has been used for experimental purpose. At optimum condition a maximum ~83 % degradation efficiency has been observed.

Recently, researchers are dealing with effluent waste water by using photoelectrocatalysis method which is a modified version of eletrocatalysis method. Paul et al. [64] showed a degradation efficiency of ~80 % for rhodamine-B dye under solar irradiation by photoelectrocatalysis method. Metal-free few-layer MoS_2 nanoflower/TiO_2 nanobelts heterostructure anode has been used in a three compartment electrochemical cell. Pt, and Ag/AgCl (saturated KCl) were used as auxiliary and reference electrodes. Boron-doped TiO_2 nanotubes synthesized by electrochemical anodization process were used to degrade acid yellow 1 under UV irradiation by Bessegato et al. [65] resulting in ~100 % discoloration. A conventional three electrode cell consisted of Ti/Ru plate auxiliary electrode and Ag/AgCl reference electrode under UV irradiation has been used for experimental purpose. Liu et al. [66] studied the discoloration (~96 %) of Rhodamine B by carbon nanotube/agarose anode with Pt auxiliary electrode and saturated calomel electrode as reference in a three compartment electrolytic cell. The anode was fabricated as thin film on indium tin oxide glass. Mumjitha and Raj [67] degraded methyl orange dye with the help of TiO_2–SiO_2 ceramic coated electrode by using photoelectrocatalysis method. Anodized electrodes were kept in a single compartment electrolytic cell with Na_2SO_4 as supporting electrolyte. The effect of voltage, initial pH, and chloride ion concentration has been studied and result showed ~100 % discoloration efficiency. Adhikari et al. [68] carried out degradation experiments by using a silver nano particle coated spherical silicon carbide electrode as anode in three compartment electrolytic cell. Pt and standard calomel electrode equipped cell produced a degradation efficiency of 80 % and 98 % for Orange G and amido black dyes, respectively. Theerthagiri et al. [69] degraded direct red 81 dye by using α-Fe_2O_3–g-C_3N_4 nanocomposites as anode, Pt-wire as counter electrode and Ag/AgCl (in saturated KCl) as a reference electrode. Result showed a degradation efficiency of ~80 % for direct red 81 dye. Brillas and Martinez-Huitle [7] investigated the use of various photocatalyst such as TiO_2, WO_3, ZnO, Bi_2WO_6, BDD-$ZnWO_4$, $BiVO_4$, $BiPO_4$, SnO, etc. for degradation of dye by photoelectrocatalytic method.

6.2.4   Electrochemical reduction method

Nowadays researchers are showing less interest in electroreduction method for degradation of dye due to its very low degradation efficiency [15]. Betchtold et al. [70] used electroreduction method to degrade highly colored pad patch dyeing effluents containing reactive dyes. The experimental set up consisted of cathodes made of stainless steel and a mixed metal oxide coated Pt anode. The removal efficiencies have been shown as, remazol black B ~14.6 %, drimaren brillant red ~9.8 % and cibachron blue ~2%. Several other dyes such as drimaren brillant yellow, cibachron orange, and levafix marine has not been degraded by this process. Further, the author has observed that degradation could be increase up to 80-90 % by direct cathodic reduction method [71]. A multi cathode electrolyser cell has been used for the experimental purpose to degrade azo dyes. Experiments have been carried out by using stainless steel electrodes. Further, Turcanu and Betchtold [72] studied the degradation of levafix amber, levafix fast red, levafix blue and the degradation efficiencies were found to be 70 %, 90 %, and 91 %, respectively. Stainless steel electrodes were used for the experimental purpose and effect of experimental parameters like current usage and energy efficiency on dye degradation have been studied. Again, Khenifi et al. [73] suggested a similar result of higher degradation efficiency. Experiments showed a degraded efficiency ~100 % for orange G dye by using a three compartment electroreduction cell. Nickel in conjunction with Pt auxiallary electrode as cathode and Ag/AgCl reference

Table 6.1 Summary of utilization of electrochemical process for effluent treatment

Method	Dye degraded	Electrodes		Degradation efficiency (%)	Reference
		Anode	Cathode		
Electrocoagulation	Acid brown 14	Fe	Fe	81.7	Bassyouni et al. [21]
Electrocoagulation	Direct brown 2, Cibacete blue	Al	Al	99.9	Ghernaout et al. [23]
Electrocoagulation	Indigo dye	Fe, Al	Ti	100	Gilpavas et al. [24]
Electrocoagulation	Reactive blue 21	Fe	Stainless steel	98	Ardhan et al. [25]
Electrocoagulation	Ramazol red 133	Steel wool	Al	99	Lopez et al. [26]
Electrocoagulation	Reactive brilliant blue	Fe	Al	96	Fu [27]
Electrocoagulation	Methylene blue	Fe	Al	97	Duan et al. [28]
Electrocoagulation	Indigo dye	Fe	Fe	93.7	Hendaoui et al. [29]
Electrocoagulation	Acid red 1	Al	Al	91	Hosiny et al. [30]
Electrocoagulation	Acid red 336	Al	Al	95	Khemila et al. [31]
Electrocoagulation	Reactive blue 19	Fe	Fe	~95	Naje et al. [32]
Electrocoagulation	Remazol yellow	Al	Al	70	Silva et al. [33]
Electrocoagulation	Reactive blue 21	Fe	Stainless steel	~99	Phalakornkule et al. [34]
Electrochemical oxidation	Real textile effluent	p-Si/BDD	Stainless steel	74	Abdessamad et al. [39]
Electrochemical oxidation	Reactive orange 16 Reactive violet 4 Reactive red 228 Reactive black	BDD	Carbon/PTFE ADE	95.5, 98.5, 74.6, 85.5	Soares et al. [41]
Electrochemical oxidation	Real textile effluent	BDD	BDD	60.8	Sanchez et al. [42]

Method	Dye degraded	Electrodes		Degradation efficiency (%)	Reference
		Anode	Cathode		
Electrochemical oxidation	Real textile effluent	Ti/TiO$_2$-NTs/Sb-SnO$_2$	Pt-Auxiliar, Ag/AgCl/KCl	52.9	Rao et al. [43]
Electrochemical oxidation	Acid red 73	Nb–Ti / Nb-TiO2-NTs / SnO$_2$-Sb	Ti	90	Xu et al. [44]
Electrochemical oxidation	Acid green 28	Ti/β-PbO$_2$ or Ti-Pt/ β-PbO$_2$	Ni	90	Irikura et al. [45]
Electrochemical oxidation	Acid yellow 23	Graphite	Ti	~99	GilPavas et al. [23]
Electrochemical oxidation	Real textile effluent	Ti/RuO$_2$	Al	97.3	Kaur et al. [46]
Electrochemical oxidation	Reactive violet 2 Acid brown 14	Graphite anode	stainless steel	95 69.8	Hamad et al. [40]
Electro Fenton process	Evens blue	BDD	Graphite air diffusion cathode	100	Antonin et al. [54]
Solar photoelectron Fenton	Evens blue	BDD	Graphite air diffusion cathode	98	Antonin et al. [54]
Electro Fenton process	Methylene blue, titan yellow	Fe	Graphite	~96–98	Ghosh et al. [55]
Solar photoelectron Fenton	Real waste water	Si/BDD	Ti	100	GilPavas et al. [56]
Electro Fenton process	Reactive black 5	Graphite	Graphite	90	Iglesias et al. [57]
Electro Fenton process	Real textile waste water	Ti/RuO$_2$	Al	74.1	Kaur et al. [46]
Electro Fenton process	Acid red 18, Real waste water	Iron	Iron	100, 90.5	Malakooti and Moridi [58]

Method	Dye degraded	Electrodes		Degradation efficiency (%)	Reference
		Anode	Cathode		
Photocatalyitc Electro Fenton process	Acid red 14 Acid blue 92	Graphite	Graphite	~65–80	Pajootan et al. [59]
Electro Fenton process	Reactive blue 19	RuO_2/IrO_2	RuO_2/IrO_2	100	Shwe et al. [60]
Photoelectrocatalysis	Rhodamine-B	MoS_2 nanoflower/ TiO_2 flower	Pt, and Ag/AgCl (saturated KCl)	~80	Paul et al. [64]
Photoelectrocatalysis	Acid yellow 1	BDD/TiO_2	Ti/Ru, Ag/AgCl	100	Bessegato et al. [65]
Photoelectrocatalysis	Rhodamine B	Carbon Nanotube/ Agarose	Pt/Saturated Colomel electrode	96	Liu et al. [66]
Photoelectrocatalysis	Methyl orange	TiO_2–SiO_2 ceramic coating	Pt/Saturated Colomel electrode	100	Mumjitha, and Raj [67]
Photoelectrocatalysis	Orange G, Amido black	Spherical Ag decorated β SiC	Pt/calomel electrode	80 98	Adhikari et al. [68]
Photoelectrocatalysis	Direct red 81	$\alpha\text{-}Fe_2O_3$–g-C_3N_4	Pt/(Ag/AgCl)	80	Theerthagiri et al. [69]
Electrochemical reduction method	Remazol black B, drimaren brillant red, cibachron blue etc.	Metal oxide coated Pt anode	Stainless steel	80–90	Betchtold et al. [70]
Electrochemical reduction method	Levafix amber, levafix fast red, levafix blue	Stainless steel	Stainless steel	70–91	Turcanu and Betchtold [72]
Electrochemical reduction method	Orange G	Ag/AgCl	Nickel Pt	100	Khenifi et al. [73]

electrode has been used for the experimental purpose. The discrepancies in color removal efficiency are attributed to the nature of chromophores present in the dye. The application of electrochemical reduction for textile effluents are currently limited to decolorize highly color intensive effluents such as pad patch dyeing effluents containing reactive dyes [7,15].

There are several other methods that have not been explored for example, electrodialysis method. Electrodialysis process use membrane separation technology which could degrade very small amount of contaminants, such as drinking water. Electrodialysis has high degradation efficiency as compared to other membrane separation techniques. Nowak et al. [13] conducted the treatment of organic dyes by electrodialysis method. The degradation efficiencies of various anionic dyes such as methylene orange, indigo carmine, amido black, titan yellow, direct green, direct blue, direct black, etc. has been investigated as ~99 %.

A summary of all the above-discussed electrochemical method, electrode used, dye used, and degradation efficiency is provided in Table 6.1.

6.3   Conclusions

In this paper, an overview of the development of electrochemical method for various toxic dye degradation in the field of textile effluent treatment has been discussed. In general, electrochemical methods such as electrocoagulation, electrocatalysis, photo-electrocatalysis, and electrochemical oxidation, electrochemical reduction, etc. have been used for dye degradation depending on the type of dye. This paper gives an idea about the utilities of electrochemical methods regarding degradation of dye such as reactive violet 2, acid brown 14, methyl blue, methyl orange and methyl red and rhodamine 6G and rhodamine B etc. Taking advantage of versatility, easy process and not producing toxic byproduct by electrochemical method, dye degradation performance of electrochemical method could be better as compared to conventional method.

References

1. Martinez-Huitle, C. A.; Brillas, E. Decontamination of wastewaters containing synthetic organic dyes by electrochemical methods. A general review. *Appl. Catal. B: Environ.* **2009**, *87*, 105–145.

2. Robinson, T.; McMullan, G.; Marchant, R.; Nigam, P. Remediation of dyes in textile effluent: A critical review on current treatment technologies with a proposed alternative. *Bioresour. Technol.* **2001**, *77*, 247–255.

3. Savin, I. I.; Butnaru, R. Wastewater characteristics in textile finishing mills. *Environ. Eng. Manage. J.* **2008**, *7*, 859–864.

4. Vaghela, S. S.; Jethva, A. D.; Mehta, D. B.; Dave, S. P.; Ramachandraiah, S. A. G. Laboratory studies of electrochemical treatment of industrial azo dye effluent. *Environ. Sci. Technol.* **2005**, *39*, 2848–2855.

5. Naim, M. M.; El Abd, Y. M. Removal and recovery of dyestuffs from dyeing wastewaters. *Sep. Purif. Methods.* **2002**, *31*, 171–228.

6. Forgacs, E.; Cserhati, T.; Oros, G. Removal of synthetic dyes from waste waters: A review. *Environ. Int.* **2004**, *7*, 953–971

7. Brillas, E.; Martinez-Huitle, C. A. Decontamination of wastewaters containing synthetic organic dyes by electrochemical methods. *Appl. Catal. B: Environ.* **2015**, *166–167*, 603–643.

8. dos Santos, A. B.; Cervantes, F. J.; van Lier, J. B. Review paper on current technologies for decolourisation of textile wastewaters: Perspectives for anaerobic biotechnology. *Bioresour. Technol.* **2007**, *98*, 2369–2385.

9. Bhatnagar, R.; Joshi, H.; Mall, I. D.; Srivastava, V. C. Electrochemical oxidation of textile industry wastewater by graphite electrodes. *J. Environ. Sci. Health A Toxic Hazard. Subst. Environ. Eng.* **2014**, *49*(8), 955–966.

10. Khenifi, A.; Bouberka, Z.; Hamani, H.; Illikti, H.; Kameche, M.; Derriche, Z. Decoloration of Orange G (OG) using electrochemical reduction. *Environ. Technol.* **2012**, *33*, 1081–1088.

11. Bayramoglu, M.; Eyvaz, M.; Kobya, M. Treatment of the textile wastewater by electrocoagulation: Economical evaluation. *Chem. Eng. J.* **2007**, *128*(2–3), 155–161.

12. Zodi, S.; Merzouk, B.; Potier, O.; Lapicque, F.; Leclerc, J. P. Direct red 81 dye removal by a continuous flow electrocoagulation/flotation reactor. *Sep. Purif. Technol.* **2013**, *108*, 215–222.

13. Nowak, K. M. Treatment of organic dye solutions by electrodialysis, The 2012 World Congress on Advances in Civil, Environmental, and Materials Research (ACEM' 12) 2012; Seoul, Korea, August 26–30.

14. Haque, M. M.; Smith, W. T.; Wong, D. K. Y. Conducting polypyrrole films as a potential tool for electrochemical treatment of azo dyes in textile wastewaters. *J. Hazard. Mater.* **2015**, *283*, 164–170.

15. Sala, M.; Gutierrez-Bouz, M. C. Electrochemical techniques in textile processes and wastewater treatment. *Int. J. Photoenergy.* **2012**, *2012*, 1–12.

16. Bensalaha, N.; Quiroz Alfarob, M. A.; Martinez-Huitlec, C. A. Electrochemical treatment of synthetic wastewaters containing Alphazurine A dye. *Chem. Eng. J.* **2009**, *149*, 348–352.

17. Chatzisymeon, E.; Xekoukoulotakis, N. P.; Coz, A.; Kalogerakis, N.; Mantzavinos, D. Electrochemical treatment of textile dyes and dye house effluents. *J. Hazard. Mater.* **2006**, *137*, 998–1007.

18. Garcia-Segura, S.; Eiband, M. M. S. G.; de Melo, J. V.; Martínez-Huitle, C. A. Electrocoagulation and advanced electrocoagulation processes: A general review about the fundamentals, emerging applications and its association with other technologies. *J. Electroanal. Chem.* **2017**, *801*, 267–299.

19. Aquino, J. M.; Pereira, G. F.; Rocha-Filho, R. C.; Bocchi, N.; Biaggio, S. R. Combined coagulation and electrochemical process to treat and detoxify a real textile effluent, *Water Air Soil Pollut.* **2016**, *227*, 266.

20. Assemian, A. S.; Kouassi, K. E.; Zogbe, A. E.; Adouby, K.; Drogui, P *In-situ* generation of effective coagulant to treat textile bio-refractory wastewater: Optimization through response surface methodology. *J. Environ. Chem. Eng.* **2018**, *6*(4), 5587–5594.

21. Bassyouni, D. G.; Hamada, H. A.; El-Ashtoukhy, E. -S. Z.; Amin, N. K.; Abd El-Latifa, M. M. Comparative performance of anodic oxidation and electrocoagulation as clean processes for electrocatalytic degradation of diazo dye Acid Brown 14 in aqueous medium. *J. Hazard. Mater.* **2017**, *335*, 178–187.

22. Assemian, A. S.; Kouassi, K. E.; Drogui, P.; Adouby, K.; Boa, D. Removal of a persistent dye in aqueous solutions by electrocoagulation process: Modeling and optimization through response surface methodology. *Water Air Soil Pollut.* **2018**, *229*, 184.

23. Ghernaout, D.; Al-Ghonamy, A. I.; Messaoudene, N. A.; Aichouni, M.; Naceur, M. W.; Benchelighem, F. Z.; Boucherit, A. Electrocoagulation of direct brown 2 (DB) and BF cibacete blue (CB) using aluminum electrodes. *J. Sep. Sci. Technol.* **2015**, *50*(9), 1413–1420.

24. GilPavas, E.; Arbeláez-Castaño, P.; Medina, J.; Acosta, D. A. Combined electrocoagulation and electro-oxidation of industrial textile wastewater treatment in a continuous multi-stage reactor electrocoagulation and electro oxidation of industrial textile wastewater treatment. *Water Sci. Technol.* **2017**, *76*(9–10), 2515–2525.

25. Ardhan, N.; Moore, E. J.; Phala, C. Novel anode made of iron scrap for a reduced-cost electrocoagulator. *Chem. Eng. J.* **2014**, *253*, 448–455.

26. Lopez, A.; Valero, D.; Cruz, L. G.; Saez, A.; Garcıa, V. G.; Exposito, E.; Montiel, V. Characterization of a new cartridge type electrocoagulation reactor (CTECR) using a three-dimensional steel wool anode. *J. Electroanal. Chem.* **2016**, *793*, 93–98.

27. Fu, Z. Treatment of reactive brilliant blue x-br synthetic wastewater by asynchronous periodic reversal electrocoagulation and its strengthening mechanism. *Water Air Soil Pollut.* **2018**, 229, 69.

28. Duan, X.; Wu, P.; Pi, K.; Zhang, H.; Liu, D.; Gerson, A. R. Application of modified electrocoagulation for efficient color removal from synthetic methylene blue wastewater. *Int. J. Electrochem. Sci.* **2018**, *13*, 5575–5588.

29. Hendaoui, K.; Ayari, F.; Rayana, I. B.; Ammar, R. B.; Darragi, F.; Trabelsi-Ayadi, M. Real indigo dyeing effluent decontamination using continuous electrocoagulation cell: Study and optimization using response surface methodology. *Process Saf. Environ. Protec.* **2018**, *116*, 578–589.

30. El-Hosiny, F. I.; Khalek, M. A. A.; Selim, K. A.; Osama, I. Physicochemical study of dye removal using electro-coagulation flotation process. *Physicochem. Prob. Miner. Process.* **2018**, *54*(2), 321–333.

31. Khemila, B.; Merzouk, B.; Chouder, A.; Zidelkhir, R.; Leclerc, J.; Lapicqu F. Removal of a textile dye using photovoltaic electrocoagulation. *Sustain. Chem. Pharm.* **2018**, *7*; 27–35.

32. Naje, A. S.; Chelliapan, S.; Zakaria, Z.; Ajeel, M. A.; Sopian, K.; Hasan, H. A. Electrocoagulation by solar energy feed for textile wastewater treatment including

mechanism and hydrogen production using a novel reactor design with a rotating anode. *R. Soc. Chem. Adv.* **2016**, *6*, 10192–10204.

33. Silva, T. B. P.; Tehuitzil, H. R.; del Río, G. M.; Villamar, J. H. Photovoltaic energy-assisted electrocoagulation of a synthetic textile effluent. *Int. J. Photoenergy.* **2018**, *2018*, 1–9.

34. Phalakornkule, C.; Luanwuthi , T.; Neragae , P.; Moore, E. J. A continuous-flow sparged packed-bed electrocoagulator for dye decolorization. *J. Taiwan Inst. Chem. Eng.* **2016**, *64*, 1–10.

35. Assemian, A. S.; Kouassi, K. E.; Drogui, P.; Adouby, K.; Boa, D. Removal of a persistent dye in aqueous solutions by electrocoagulation process: Modeling and optimization through response surface methodology. *Water Air Soil Pollut.* **2018**, *229*, 184.

36. Nidheesh, P.V.; Zhou, M.; Oturan, M.A. An overview on the removal of synthetic dyes from water by electrochemical advanced oxidation processes. *Chemosphere.* **2018**, *197*, 201–227.

37. Ganzenko, O.; Huguenot, D.; Hullebusch, E. D.; Esposito G.; Oturan, M. A. Electrochemical advanced oxidation and biological processes for wastewater treatment: A review of the combined approaches. *Environ. Sci. Pollut. Res.* **2014**, *21*(14), 8493–8524.

38. Sires, I.; Brillas, E.; Oturan, M. A.; Rodrigo, M. A.; Panizza, M. Electrochemical advanced oxidation processes: Today and tomorrow. A review. *Environ. Sci. Pollut. Res.* **2014**, *21*(14), 8336–8367.

39. Abdessamad, N. E. H.; Akrout, H.; Bousselmi, L. Anodic oxidation of textile wastewaters on boron doped diamond electrodes. *Environ. Technol.* **2015**, *36*(24), 3201–3209.

40. Hamad, H.; Bassyouni, D.; Ashtoukhy, E. S. E.; Amin, N.; Latif, M. A. E. Electrocatalytic degradation and minimization of specific energy consumption of synthetic azo dye from wastewater by anodic oxidation process with an emphasis on enhancing economic efficiency and reaction mechanism. *Ecotoxicol. Environ. Saf.* **2018**, *148*, 501–512.

41. Soares, I. C. D.C.; Silva, D. R. D.; Nascimento, J. H. O. D.; Segura, S. G.; Martinez-Huitle, C. A. Functional group influences on the reactive azo dye decolorization performance by electrochemical oxidation and electro-Fenton technologies. *Environ. Sci. Pollut. Res.* **2017**, *24*(31), 24167–24176.

42. Sanchez, A. S.; Perez, M. T.; Rivas, R. M. F.; Hernandez, I. L.; Miranda, V. M.; Fonseca-M.D Oca, R. M. G. Treatment of a textile effluent by electrochemical oxidation and coupled system electooxidation-salix babylonica. *Int. J. Photoenergy.* **2018**, *2018*, 1–12.

43. Rao, A. N. S.; Venkatarangaiah, V. T.; Nagarajappa, G. B.; Nataraj, S. H.; Krishnegowd, P. M. Enhancement in the photo-electrocatalytic activity of SnO_2–Sb_2O_4 mixed metal oxide anode by nano-WO_3 modification: Application to trypan blue dye degradation. *J. Environ. Chem. Eng.* **2017**, *5*, 4969–4979.

44. Xu, L.; Liang, G.; Yin, M. A promising electrode material modified by Nb-dopedTiO$_2$ nanotubesfor electrochemical degradation of AR 73. *Chemosphere.* **2017**, *173*, 425–434.

45. Irikura, K.; Bocchi, N.; Rocha-Filho, R. C.; Biaggio, S. R.; Iniesta, J.; Montiel, V. Electrodegradation of the acid green 28 dye using Ti/β-PbO$_2$ and Ti-Pt/β-PbO$_2$ anodes. *J. Environ. Manage.* **2016**, *183*, 306–313.

46. Gilpvaz, E.; Dobrosz-Gomezb, I.; Gomez-Garciac, M. A. Optimization of solar-driven photo-electro-Fenton process for the treatment of textile industrial wastewater. J. *Water Process Eng.* **2018**, *24*, 49–55.

47. Kaur, P.; Kushwaha, J. P.; Sangal, V. K. Transformation products and degradation pathway of textile industry wastewater pollutants in electro-Fenton process. *Chemosphere.* **2018**, *207*, 690–698.

48. Ferreira, M. B.; Rocha , J. H. B.; DeSilva, D. R.; De Moura, D. C.; De Araujo, D. M.; Martinez-Huitle, C. A. Application of electrochemical oxidation process to the degradation of the Novacron Blue dye using single and dual flow cells. *J. Solid State Electrochem.* **2017**, *24*(31), 24167–24176.

49. Anglada, A.; Urtiaga, A.; Orti, I. Contributions of electrochemical oxidation to waste-water treatment: Fundamentals and review of applications. *J. Chem. Technol. Biotechnol.* **2009**, *84*, 1747–1755.

50. Rodriguez, F. A.; Rivero, E. P.; Gonzalez, I. Electrogeneration of active chlorine in a filter-press-type reactor using a new SB$_2$O$_5$ doped Ti/RUO$_2$-ZRO$_2$ electrode: Indirect indigoid dye oxidation. *Int. J. Chem. Reactor Eng.* **2017**, *15*, 0095.

51. Rajkumar, D.; Kim, J. K. Oxidation of various reactive dyes with in situ electro-generated active chlorine for textile dyeing industry wastewater treatment. *J. Hazard. Mater.* **2006**, *136*, 203–212.

52. Rajkumar, K.; Muthukumar, M. Optimization of electro-oxidation process for the treatment of Reactive Orange 107 using response surface methodology. *Environ. Sci. Pollut. Res.* **2012**, *19*, 148–160.

53. Ma, J.; Song, W.; Chen, C.; Ma, W.; Zhao, J.; Tang, Y. Fenton degradation of organic compounds promoted by dyes under visible irradiation. *Environ. Sci. Technol.* **2005**, *39*, 5810–5815.

54. Antonin, V. S.; Garcia-Segura, S.; Mouria, C.; Brillas, E. Degradation of Evans blue diazo dye by electrochemical processes based on Fenton's reaction chemistry. *J. Electroanal. Chem.* **2015**, *747*, 1–11.

55. Ghosh, P.; Thakur, L. K.; Samanta, A. N.; Ray, S. Electro-Fenton treatment of synthetic organic dyes: Influence of operational parameters and kinetic study. *Korean J. Chem. Eng.* **2012**, *29*(9), 1203–1210.

56. Gilpavas, E.; Dobrosz-Gomez, I.; Gomez-Garcia, M. A. Optimization of solar-driven photo-electro-Fenton process for the treatment of textile industrial wastewater. *J. Water Process Eng.* **2018**, *24*, 49–55.

57. Iglesias, O.; de Dios, M. A. F.; Rosales, E.; Pazos, M.; Sanroman, M. A. Optimisation of decolourisation and degradation of reactive black 5 dye under electro-Fenton process using Fe alginate gel beads. *Environ. Sci. Pollut. Res.* **2013**, *20*, 2172–2183.

58. Malakootian, M.; Moridi, A. Efficiency of electro-Fenton process in removing Acid Red 18 dye from aqueous solutions. *Process Saf. Environ. Protec.* **2017**, *111*, 138–147.

59. Pajootan, E.; Arami, M.; Rahimdokht, M. Application of carbon nanotubes coated electrodes and immobilized tio$_2$ for dye degradation in a continuous photocatalytic-electro-Fenton process. *Ind. Eng. Chem. Res.* **2014**, *53*(42), 16261–16269.

60. Shwe, M. T.; Dimaculangan, M. M.; De Luna, M. D. G. Decolourization of simulated dye wastewater containing reactive blue 19 (RB19) by the electro-Fenton reaction in the presence of metal oxide-coated electrodes. *Adv. Mater. Res.* **2014**, *858*, 40–45.

61. Urmi, S. A.; Khurny, A. S. W.; Gulshan, F. Decolorisation of methyl orange using mill scales by photo electro Fenton process. *Proc. Eng.* **2015**, *105*, 844–851.

62. Zhang, Y.; He, P.; Jia, L.; Li, C.; Liu, H.; Wang, S.; Zhou, S.; Dong, F. Ti/PbO$_2$–Sm$_2$O$_3$ composite based electrode for highly efficient electrocatalytic degradation of alizarin yellow R. *J. Colloid Interface Sci.* **2019**, *533*, 750–761.

63. He, Y. Z. P.; Ji, L.; Li, C.; Liu, H.; Wang, S.; Zhou, S.; Dong, F. Ti/PbO$_2$–Sm$_2$O$_3$ composite based electrode for highly efficient electrocatalytic degradation of alizarin yellow R. *J. Colloid Interface Sci.* **2019**, *533*, 750–761.

64. Paul, K. K.; Sreekanth, N.; Biroju, R. K.; Narayanan, T. N.; Giri, P. K. Solar light driven photoelectrocatalytic hydrogen evolution and dye degradation by metal-free few-layer MoS$_2$ nanoflower/TiO$_2$(B) nano beltshetero structure. *Solar Energy Mater. Solar Cells.* **2018**, *185*, 364–374.

65. Bessegato, G. G.; Cardoso, J. C.; Zanoni, M. V. B. Enhanced photoelectrocatalytic degradation of an acid dye with boron-doped TiO$_2$ nanotube anodes. *Catal. Today* **2015**, *240*, 100–106.

66. Liu, H.; Ren, M.; Zhang, Z.; Qu, J.; Ma, Y.; Lu, N. A novel electrocatalytic approach for effective degradation of Rh-B in water using carbon nanotubes and agarose. *Environ. Sci. Pollut. Res.* **2018**, *25*, 12361–12372.

67. Mumjitha, M.; Raj, V. Electrochemical synthesis, structural features and photoelectrocatalytic activity of TiO$_2$–SiO$_2$ ceramic coatings on dye degradation. *Mater. Sci. Eng. B* **2015**, *198*, 62–73.

68. Adhikari, S.; Eswar, N. K.; Sangita, S.; Sarkar, D.; Madras, G. Investigation of nano Ag-decorated SiC particles for photoelectrocatalytic dye degradation and bacterial inactivation. *J. Photochem. Photobiol. A: Chem.* **2018**, *357*, 118–131.

69. Theerthagiri, J.; Senthil, R. A.; Priya, A.; Madhavan, J.; Michael, R. J. V.; Ashokkumar, M. Photocatalytic and photoelectrochemical studies of visible-light active α-Fe$_2$O$_3$–g-C$_3$N$_4$ nanocomposite. *RSC Adv.* **2014**, *4*, 38222–38229.

70. Bechtold, T.; Burtscher, E.; Turcanu, A. Cathodic decolourisation of textile waste water containing reactive dyes using a multi-cathode electrolyser. *J. Chem. Technol. Biotechnol.* **2001**, *76*, 303–311.

71. Bechtold, T.; Mader, C.; Mader, J. Cathodic decolourization of textile dyebaths: Tests with full scale plant. *J. Appl. Electrochem.* **2002**, *32*, 943–950.

72. Turcanu, A.; Bechtold, T. Cathodic decolourisation of reactive dyes in model effluents released from textile dyeing. *J. Clean. Prod.* **2017**, *142*, 1397–1405.

73. Khenifi, A.; Bouberka, Z.; Hamani, H.; Illikti, H.; Kameche, M.; Derriche, Z. Decoloration of Orange G (OG) using electrochemical reduction. *Environ. Technol.* **2012**, *33*, 1081–1088.

7

Reuse and recycling of water for sustainable textile chemical processing

Sanjay Kumar Bhikari Charan Panda,

Samrat Mukhopadhyay, Javed Sheikh*

Dept. of Textile and Fibre Engineering, Indian Institute of Technology (IIT),

Delhi, India

*Corresponding Author. E-mail: jnsheikh@textile.iitd.ac.in

Abstract: Textile industries consume plenty of water and also produce a significant amount of effluent. The effluents are rich in color, contain a high concentration of organic compounds, and are with a high level of pH, COD, BOD, TOC, TDS, etc. Researchers are in continuous experimentation for many advanced treatment processes to treat such contaminated water. Advanced treatments like electrochemical- treatments, catalytic-ozonation, ultrasonic-assisted ozone-oxidation, photocatalytic-oxidation with UV/H_2O_2, Fenton/Photo-Fenton oxidation, etc. are the outputs and most of these treatments are confined to laboratory-scale until now. The conventional bio chemical process removes up to 90% of organic pollutants from wastewater, while the removal of COD and TDS is challenging. The best alternative to this is the use of reverse osmosis, which is supposed to be a costly process. Hence to tackle the scarcity of water, the adoption of clean technology has become an absolute necessity, which involves a reduction in water usage and reuse of water. In this context, the direct reuse of wastewater without any treatment within the system will be most beneficial. The present paper reviews various water management techniques in textile chemical processing with the main focus on direct reuse of process water.

Keywords: Textile Wastewater, Reuse, Recycle, Sustainability, Effluent Treatment.

7.1   Introduction

Water is an abundant resource. It is a well-known fact that the sea covers almost three fourth of the earth's surface. However, out of the total volume of water available on earth, only 2.70% is freshwater. Out of the majority of freshwater, 75.20% is in the form of ice in the pole region, while 22.60% is groundwater. The rest of the water is available in the lake, river, atmosphere, etc. which is very small in volume [24].

Textile industries consume a massive quantity of water and also produce a significant amount of effluent during the production processes like desizing, scouring, bleaching, dyeing, and finishing. All wet processing treatments are water-intensive and generate a high volume of effluent containing a significant quantity of unexhausted dye, high concentration of organic compounds, and with a high level of pH, COD, BOD, TOC, TDS, etc. A few examples here show to what extent the largest textile exporters are consuming water. China is one of the biggest textile exporters in the world which has consumed 4 billion cubic meters of water in the year 2014. The second exporter, European Union water usage, was 600 million cubic meters. Turkey shows water consumption in a range of 50–100 cubic meters per ton of finished textiles produced [5].

The textile industry is the largest industry in the Indian economy, which contributes almost 15% of the total exports. It is one such labor-intensive industry that employs 51 million people directly and 68million people indirectly. During the financial year 2015–2016, the overall textiles export was US$ 40 billion. The market growth is to be expected to reach approximately US$ 230 billion by 2020. The textile industry in India contributes 4% of the gross domestic products (GDP) and 14% to an overall index of industrial pollution (IIP) [16].

7.2   Textile process and source of effluent

The textile industry is involved with various mechanical and chemical processes during the conversion of fabric from the fiber. There are mainly two types of processes called dry process and wet process. The dry process includes opening, mixing, blending, carding, combing, spinning, weaving, and knitting which consumes a minimal amount of water. The wet process includes desizing, scouring, mercerizing, bleaching, dyeing and finishing which consumes a lot of water and produces a significant volume of effluent.

Desizing is a process where the size material applied before weaving is removed to facilitate further processes. Scouring is the treatment of cotton textiles with alkaline liquor at boiling temperature to make it hydrophilic and bleaching removes the natural coloring matters. Mercerization is a treatment of cotton yarn or fabric with cold and highly concentrated (270–300 g L^{-1}) sodium hydroxide solution to improve dimensional stability, luster, strength, and dyeability. Dyeing and printing are the processes to color the textiles and finishing add value to textiles in different ways. The different types of chemicals used in the above-said processes are the pollutants that pollute the influent at a different level. Each of these processes mentioned generates the effluent in massive quantity with varying characteristics, as specified in Table 7.1.

Indian textile industries consume a massive quantity of raw water in wet processing, which is in the range of 61–646 L kg^{-1} of fabric processed. The average wastewater generated is 172 L kg^{-1} of fabric processed. The

Table 7.1   Types of pollutant and their origin in textile wastewater

Sr. No.	Pollutants	Chemical type	Source
1	Organic load	Starches, enzymes, surfactants, fats, greases, waxes, acetic acid	Desizing, scouring, bleaching, dyeing
2	Color	Dyes	Dyeing
3	Nutrients (N and P)	Ammonium salts, urea phosphate-based buffers, sequestrants	Dyeing
4	pH and salt effects	Sodium hydroxide, mineral/organic acids, sodium chloride, silicates, carbonates	Desizing, scouring, bleaching, mercerizing, dyeing
5	Sulfur	Sulfates, sulfides, hydrosulfite salts and sulfuric acids	Dyeing
6	Toxicants	Heavy metals, reducing agents (sulfides), oxidizing agents (chlorites, peroxides, dichromates, persulfates), biocides, quaternary ammonium salts	Desizing, bleaching, dyeing, finishing
7	Refractory organics	Surfactants, dyes, resins, synthetic size (PVA), chlorinated organic compounds, carrier organic solvents	Desizing, bleaching, dyeing, finishing

wastewater generated is with different compositions depends on the type of fabric used, the process adopted, and the types of machines utilized for processing. The characteristic of the textile effluent is shown in Table 7.2 [26].

Table 7.2   Characteristic of textile effluent

Sr. no.	Process	Effluent composition	Nature
1	Sizing	Starches, polyvinyl alcohol (PVA), carboxymethylcellulose (CMC), wetting agent	High in BOD, COD
2	Desizing, scouring	Starches, PVA, CMC, fat, waxes, pectin	High in BOD, COD, SS, DS
3	Bleaching	Sodium hydroxide, sodium hypochlorite, hydrogen peroxide, sodium silicate, sodium phosphates, cotton fiber	High in alkalinity, SS
4	Mercerizing	Sodium hydroxide, cotton wax	High in pH, DS
5	Dyeing	Dyes, urea, oxidizing agents, reducing agents, acids, wetting agents, detergents	High in color, BOD, DS, low SS, low heavy metals
6	Printing	Paste, urea, binder, thickener, cross-linking agents, urea, starch, gum, reducing agents, alkali	High in color, BOD, SS, oily appearance

Table 7.3 Characteristics of textile effluents from textile chemical processing

Sr. no.	Characteristics	Scouring	Bleaching	Mercerizing	Dyeing
1	pH	10–12	8.5–11	8–10	9–11
2	TDS (mg L^{-1})	12000–30000	2500–11000	2000–2600	1500–4000
3	TSS (mg L^{-1})	1000–2000	200–400	100–400	50–350
4	BOD (mg L^{-1})	2500–3500	100–500	50–120	100–400
5	COD (mg L^{-1})	10000–20000	1200–1600	250–400	400–1400
6	Chloride (mg L^{-1})	–	–	350–700	–
7	Sulphates (mg L^{-1})	–	–	100–350	–
8	Color	–	–	Highly colored	Strongly colored

Table 7.4 Effluent characteristics against discharge limits

Sr. no.	Characteristics	Composite effluent	Discharge limit
1	pH	8–10	5.6–9
2	TDS (mg L^{-1})	5000–10000	2100
3	TSS (mg L^{-1})	100–700	100
4	BOD (mg L^{-1})	50–550	30
5	COD (mg L^{-1})	250–8000	250
6	Chloride (mg L^{-1})	100–500	1000
7	Sulphates (mg L^{-1})	50–300	1000
8	Color	Strongly colored	Colorless

A typical effluent characteristic as prescribed by statutory authorities from various textile wet-processing is shown in Table 7.3.

The characteristic of the composite effluents against discharge limits, according to the Bureau of Indian Standards, is shown in Table 7.4 [20].

7.3   Effluent treatment processes for textile wastewater

The main processes used in the treatment of textile wastewater are prelimi-nary treatment, primary treatment, secondary treatment, and tertiary and/or advanced treatment.

7.3.1   Preliminary treatment

Preliminary treatments are always carried out before primary treatment in CETP (common effluent treatment plant) plants or own ETP (effluent treatment plant) plants. The intention is to separate coarse solids by using screens or grates. Oil and grease removal along with pH control is also part of this treatment.

7.3.2   Primary treatment

The primary treatment involves different methods like fine screening, coagulation, sedimentation, flocculation, floatation, etc., followed by a primary clarifier. This process removes the total suspended solids (TSS) to the order of 50–70% along with biochemical oxygen demand 25–50%. A significant portion of oil and grease (about 65%) also get removed in this treatment.

7.3.3   Secondary treatment

The suspended and dissolved organic matters are decomposed with the help of microbes in this treatment. Microbes can be used in different ways. The process in which microbes are developed in the presence of air called an aerobic process and in the absence of air called the anaerobic process and with or without air called facultative process. The microbes consume organic matters as a source of carbon and energy. This treatment is mostly used to meet the standard for the biochemical oxygen demand before getting discharged to surface water.

7.3.4   Tertiary and/or advanced treatments

Tertiary treatments are usually an extension to secondary treatments that decides the quality of final processed wastewater for reuse and disposal with accepted norms. This process involves adsorption with activated carbon, membrane filtration, ion exchange process, reverse osmosis, etc. [12].

7.4   Adsorption

Adsorption is a physicochemical process that is used in the treatment of textile wastewater mainly to remove the color. Color is an important pollutant which not only reduces the aesthetic value of water but also affects flora and fauna by inhibiting sunlight to the water stream. Activated carbon widely used as commercial adsorbent due to its high surface area. Although activated carbon is successful in the application for removal of the color from wastewater, the high material cost and cost of regeneration has forced researchers to search for viable alternatives. The other adsorbents recently tried are sugar cane bagasse, tamarind fruit shell, fly ash and red mud, fly ash and soil, lignite coal, rice husk, jack fruit peel, orange peel, wheat straw, sunflower stalks [17], sawdust [19], rice straw, corn straw, coconut husk, cotton stalks, oil palm shells, tobacco stem, soybean oil cake, corncob, lignin, wool, durean peel, sludge, palm oil ash, cocoa, papaya seed, grass waste, guava leaves, pineapple stem, coffee husk, peanut shell, neem leaf powder [1], Prickly pear peels (CarTunaQ), broccoli stems (CarbocQ), white sapote seeds (CarZapQ), etc. [27].

7.4.1   Adsorption with rice husk

Rice husk templated water treatment sludge is an excellent adsorbent for dye and metal ion removal from effluents, 5–15 (wt. %) rice husk molded along with water treatment sludge and dried one week in the sun followed by further drying at 95 °C for 24 h. Finally, it is fired in a muffle furnace at 900 °C for 3 h at a gradient of 5 °C min^{-1} to obtain a mesoporous sludge. The combustion creates several pores due to the removal of organic components results in increasing the volume per unit mass. Such mesoporous sludge is very useful to remove substances like rosaniline dye and ions like Pb^{2+}, Ni^{2+} and chlorine [33]. A geopolymer prepared using metakaolin, rice-husk-ash, and soybean oil. This geopolymer used at a concentration of 1.5 g L^{-1} to adsorb 70.8% of 50 ppm methyl violet 10B dye. The pH maintained at 4.5, and the equilibrium needs 120 min [2]. Modified rice bran is used as a low-cost adsorbent to remove 98.2% of reactive blue 4 (35%) with a dose of 65.36 mg for 500 ppm dye concentration [15]. Rice-husk-charcoal is successfully studied in Congo red, where 20 ppm Congo red solution is mixed with rice husk charcoal at 30 °C and pH 7.4 by stirring and centrifuged. The concentration of rice husk charcoal used 1.2 g L^{-1} is most effective to adsorb 20 ppm Congo red solution in 220 min and 0.6 g L^{-1} concentration shows absorbance less than 0.2 [17].

7.4.2   Adsorption with sawdust

Sawdust is also studied for its dye adsorption and COD removal efficacy on textile wastewater and found effective in the removal of COD up to 76% along with excellent dye adsorption [19]. Eucalyptus wood sawdust (*Eucalyptus globules*) is prepared by treating with 1% NaOH at 50 °C for 4 h to remove the lignin followed by washing and drying. The adsorption capacity of this sawdust is observed increasing from 5.10 mg L^{-1} to 66.67 mg L^{-1} with an increase in temperature from 288 K to 318 K, respectively [23]. A low-cost sodium hydroxide treated sawdust (*E. globules*) is also found effective adsorbent for the removal of brilliant green dye (Basic green 4). It can be used at a maximum concentration of 4 g L$^{-1,}$ where the rate of adsorption decreases with an increase in temperature [22].

7.4.3   Adsorption with agricultural by-products

Materials that are cheaper, rich in carbon content with low organics, can be used as adsorbents. Many agricultural by-products are used for this purpose and due to the availability at the cheaper price, these can be rejected after their use. There is no need for regeneration. Agricultural by-products used for activated carbon are rice straw, corn straw, sawdust, wheat straw, coconut husk, rice husk, cotton stalks, oil palm shells, tobacco stem, soybean oil cake, corncob, lignin, wool, durean peel, sludge, palm oil ash, cocoa, papaya seed, grass waste, guava leaves, pineapple stem, coffee husk, peanut shell, neem leaf powder, etc. [1].

7.4.4   Adsorption with vegetable residues

Activated carbons are available in different forms like powder (PAC), granule (GAC), fiber (ACF), cloth (ACFC), carbon nanotubes (CNT), monoliths, composites, extruded or pellet form depending on the method of preparation and precursor used. Activated carbons prepared from vegetable residues were studied for removal of dyes from wastewater and was found very effective. Prickly pear peels (CarTunaQ), broccoli stems (CarbocQ), white sapote seeds (CarZapQ) were used as the precursor for granule activated carbon prepared by carbonizing at 673 K in a muffle furnace in the presence of H_3PO_4. The activated carbon prepared from prickly pear peels, broccoli stems, and white sapote seeds shows with high specific surface area ($1025–1177$ m^2 g^{-1}), pore size ($4.1–8.4$ nm), and with pore volume ($1.06–2.16$ cm^3 g^{-1}). This study showed 93–98% removal of reactive dyes from wastewater against commercial activated carbon with 95–98% of removal. Vat dyes were removed to the extent of 53–77% against 50–70% by commercial activated carbon. The study also showed effective removal of acid, basic, and direct dyes from their solutions using adsorption on the prepared activated carbon granule [27].

7.4.5   Adsorption with multiwall carbon nanotubes

Multiwall carbon nanotubes (MWCNTs) were studied as an adsorbent for adsorption of Bismarck Brown R IUPAC name (4-[5-(2,4-diamino-5-5-methyl phenyl)diazenyl-2-methylphenyl]diaznyl-6-methylbenzol-1,3-diamin). The solubility of the dye is 11 g L^{-1} at 25 °C temperature and the molecular weight is 461.39 g mol^{-1}. This study shows for 1×10^{-5} M to 10×10^{-5} M concentration of dye solution by using the optimized quantity of MWCNTs as 25 mg the adsorption was up to 90%. The adsorption increased with an increase in pH from 2 to 5 but showed a decrease in adsorption above pH 5, concluding pH sensitivity of the process. The maximum contact time observed was 60 min whereas the maximum adsorption was observed in an initial 10 min. The adsorption rate increased with an increase in temperature [18].

Unpurified multi-walled carbon nanotubes are also found useful in adsorption of dye from wastewater at a lower cost of production due to their unique structural characteristics. The kinetic analysis shows the pseudo-second-order model is the best fitting for this study. About 8 g L^{-1} CNTs were found capable of removing dye from wastewater with a dye concentration of 50–250 mg L^{-1}. Various surface modifications with residual catalyst and oxidation showed no change in adsorption value for this study [38].

7.4.6   Adsorption with graphene sand composites

Biosynthesized graphene sand composites (GSC) were studied for its adsorption capacity, photocatalytic activity, and antibacterial activity in the textile wastewater. Graphene is an atom-like thick material with high carrier mobility ($20,000$ cm^2 v^{-1} s^{-1}) and having a large surface area ($26,000$ m^2 g^{-1}). Graphene has strong adsorption capacity. When it is used as a composite with TiO_2, Ag_3PO_4, CeO_2, and ZnO, its photocatalytic activity is enhanced. Graphene and graphene oxides were reported as very good adsorbents for pesticides, heavy metal ions, and natural dyes as well. Graphene biosynthesized from common sugar and coated with river sand were studied for the treatment of textile wastewater. It has shown very high adsorption capacity along with perfect photocatalytic activity along with high antibacterial activity against bacterium *Escherichia Coli* [28].

7.4.7   Adsorption with chitosan immobilized manganese peroxide

Worldwide 7×10^7 tons of dyes are manufactured annually, and during their application in industries, a significant amount of dyes released to the water bodies, which ultimately disturbs the ecosystem. Chitosan immobilized manganese peroxide enzyme (CI-MnP) was found to be an excellent adsorbent for the removal of dyes from textile wastewater. Chemically chitosan is composed of randomly distributed – (1, 4)-linked-glucosamine and N-acetyl-D-glucosamine units. It is an amino polysaccharide and a product of the deacetylation of chitin. It is mostly used for immobilization of enzymes due to its non-toxic nature. The benefits of using CI-MnP are color removal above 95%, 80–90% reduction in COD, TOC, BOD, and a significant decrease in cytotoxicity and mutagenicity [4].

7.4.8   Adsorption with nanosilver embedded graphene oxide with *Moringa oleifera*

Moringa oleifera seeds powder has long been used as a natural coagulant due to its low cost and is used to remove turbidity from drinking water. The proteins present in it helps in coagulation. *Moringa oleifera* also consists of benzyl isothiocyanate and benzyl glucosinolate which can assist in reducing the bacteria. Graphene oxide was reduced with aqueous extract of *Amaranthus polygonoides* and embedded with silver nanoparticles. Silver nanoparticles have excellent antibacterial properties. The nanosilver embedded graphene oxide reduced with *A. polygonoides* along with pulverized *M. oleifera* seeds was studied for its excellent adsorption capacity with antibacterial properties and it was found to give up to 98% of adsorption on synthetic effluent [9].

7.4.9   Adsorption with chitin/lignin biosorbent

Chitin/lignin (1:1) biosorbent was found very useful in the removal of C.I. Direct Blue 71 from dye solution and wastewater due to their porous structure. α-Chitin combined with kraft lignin in the presence of 15% H_2O_2 is dried at 105 °C for 12 h and passed through 100 μm sieve. The biosorbent is valid within the pH range of 2.4–8.4; however, regeneration is a drawback of the process with typical reagents like HCl, NaOH or salt. Few researchers have found the regeneration with microorganisms is quite useful [39].

7.4.10   Adsorption using polyacrylic acid hydrogel in hydrodynamic cavitations

When ultrasound energy is applied inside the water to form the cavity, there is a generation of bubbles. The bubbles grow and collapse. This process is highly exothermic. It generates high temperature (5000 K) and high pressure (1000 atm), which can easily dissociate the water to H* and OH* radicals. When such energy is applied to wastewater, there is complete mineralization of organic compounds takes place. This hydrodynamic cavitation (HC) process was studied to treat wastewater accompanied by crystal violet dye, which showed 20% decolorization and 17% removal of TOC. When this HC processed wastewater was passed through a column of polyacrylic acid hydrogel beads cross-linked with modified bentonite clay, the dye adsorption was observed up to 98% along with 70% removal of TOC [3].

7.5   Application of plasma for degradation of textile effluent

The textile industry is one of the major polluters as data shows the use of 7×10^5 tons of dyes per annum, where 10–15% of these discharged to the effluent during processing. To use ecolabels and follow norms for the discharge of effluent by the local authority, reuse of process water is an essential requirement. The electrical discharge in humid air produces many chemically active species like O^*, HO^*, N^*, HO_2^*, N_2^*, N^*, OH^-, O_2^-, O^-, O_2^+, N_2^+, N^+, O^+. These species are used to decompose organic compounds. A study was carried out on the application of coaxial dielectric barrier discharge (DBD) for the removal of reactive dyes. DBD is a typical non-equilibrium high-pressure gas discharge where AC high voltage used to provide high electron density with (1–10) eV of energy. Reactive azo dyes were selected in this study as most of these dyes are suspected as toxic, mutagenic and carcinogenic. DBD with 45 kJ L^{-1} energy discolors the colored effluent to an extent of 40–60% in 5 min and 70–97% of in 24 h. Degradation was found to be enhanced with addition of 10 mM hydrogen peroxide or iron ion as a catalyst [7].

Another study was reported for the degradation of textile wastewater using Gliding arc discharge (GAD). GAD assisted with Degussa P25 TiO_2 (80% anatase and 20% rutile) with specific area 50 m^2 g^{-1} as photocatalyst is very useful in the degradation of organic compounds. In GAD, an electric arc is established between two diverging electrodes at high voltage. The arc is pushed away from the ignition point by gas feeding, which sweeps through the electrodes gap forms a plasma plume. The system was designed in such a way that the plasma plume touches the effluent for better results. Within 30–45 min contact time for the BOD to COD ratio was above 0.7 indicating biodegradability of the waste [10].

7.6   Zero liquid discharge (ZLD)

Treatment of textile effluent, especially to reduce the total dissolved salt (TDS) is a tedious job before releasing to surface water. Even using the reverse osmosis process, which separates the TDS to the maximum extent, the saline rejects needs to be either diluted or processed again in another plant. Here the concept zero liquid discharge (ZLD) comes to picture. ZLD process separates the waste from wastewater where the water is reused and waste in the form of solid or semisolid is taken to landfill [29]. The main aim of ZLD is to recycle and reuse the waste water completely. There will be no discharge of water leaving the plant or facility boundary. The necessary process used for this system is first to pass the effluent through pressure-driven membrane filtration. Secondly, the concentrate from the first process is to be processed through membrane distillation. Finally, the waste generated from the second process must be incinerated. A study was conducted using the membrane filtration with different filters like ultrafiltration, nanofiltration with loose filters, and nanofiltration with tight filters and RO with molecular cut off in Da. 5000, 1000, 400, and 300, respectively. All four filtration processes were capable of removing dyes above 99%. COD removal was inadequate in the first two cases. RO showed the best results but the operating cost is generally very high. The benefit to cost ratio was 70–90% [37].

ZLD is not so simple to adopt due to its intensive energy consumption and very high cost. In a conventional system, the effluent treated with thermal treatment via a brine concentrator and brine crystallizer or using evaporation pond, a pre-treatment process is essential to reduce scaling potential. In brine concentrator mechanical vapor compression (MVC), multi-effect distillation (MED) or multi-stage flash (MSF) is used. This process uses approximately 52–66 kW h m^{-3} of energy. Thermal ZLD, incorporated with RO, reduces energy consumption, and is more effective. RO is also integrated with electrodialysis (ED), electrodialysis reversal (EDR), or forward osmosis (FO) to give a better result, minimizes fouling, and scaling. RO consumes 2 kW h m^{-3} of energy but the upper limit of salinity concentration is 70,000 mg L^{-1}. When it is incorporated with EDR, the level of salinity can go beyond 180,000 mg L^{-1}. Due to high osmotic pressure, it is operated at high salinity concentration. However, in FO concentrated draw solution, NH_3/CO_2 was used instead of NaCl and $MgSO_4$ [35].

7.7   Recycling of water from textile effluent and their use

7.7.1   Closed-loop recycle system

In closed-loop recycle system a specific process is looped from start to end for the water reuse from the system either directly or after some treatment. This system is viable for large textile processing industries where different processes are used with large volumes. Researchers reported a closed-loop to recycle system separately for wool scouring, cotton desizing, and wool dyeing. In wool scouring, desuinting was done first using water from a rinsing bath after scouring. The effluent from desuinting bath was collected and taken for evaporation where condensed water used as a fresh intake, and the residue was used as fertilizer. From the wool scouring bath, the detergent and wool grease were separated by ultrafiltration. The detergent was reused in the system directly and wool grease concentrate was taken for wool grease recovery. Cotton desizing effluent was processed through ultra-filtration, where the recovered size was taken for reuse and permeates recycled back to the system. Effluent generated from wool dyeing processed with treatments like ion exchange, electrocoagulation, sedimentation, and filtration. This type of processing of effluent resulted in water-saving with chemical recovery for reuse [36].

7.7.2   A scientific approach to wastewater recovery and reuse

A thorough study was conducted on a knitted fabric processing plant with an analysis of its effluent generated in each step, which resulted in the reuse of 22% of the water from total effluent generated by simple ozonation process. Water consumption in the textile knit processing industry is $20–100$ m^3 per ton of fabric processed and it depends on the process type according to the different quality of fabrics. For an average production of 10 tons per day in this plant, the wastewater generated was 751 m^3 per day. The overall wastewater generated characterized as with TDS of 9 g L^{-1}, COD of 1180 mg L^{-1}, and color value of 720 Pt-Co units. Out of the total effluent generated 750 m^3, 410 m^3 was segregated as frequently used processes. Again by stepwise observation, 127 m^3 of effluent was found with COD 350 mg L^{-1}, TDS 1.05 g L^{-1}, and color value of 25 Pt-Co units, which neither is required any expensive treatment for reuse nor such facility. After a simple ozonation process by feeding $27.65–38.10$ mg min^{-1} with contact time $15–120$ min the COD dropped up to 95 mg L^{-1} and with complete removal of color for reuse purpose [25].

7.7.3   Reuse of wastewater after nanofiltration

Nanofiltration is one of the suitable processes to retain dissolved molecules in nano size. The capability of the filter media depends upon the pore size and the charge interaction of the media. Suitable charge on the media retains the ions even smaller than their pore size. Effluent from direct dyeing, reactive dyeing, and disperse dyeing collected separately and studied for their reuse after nanofiltration. The media filter used in this study was capable of retaining molecules of size ca. 1 nm corresponds to atomic weight ca. 200. Also, the filter was negatively charged which was capable of retaining negative ions. The study showed that permeate from disperse dyed effluent after nanofiltration used in new disperse dyeing bath resulted in no deviation in dyeing and its fastness properties. Permeate from reactive dyed effluent was used in the rinsing bath of new reactive dyeing and showed poor results where sodium chloride was used. However, results were better in reactive dyeing where sodium sulphate was used. In direct dyeing, permeate used with process water at a ratio of 75:25 showed a better result in the case of sodium sulphate. The result was such as the filter media efficiency for NaCl was 10–25%, while in the case of Na_2SO_4, it was 85–95% [6].

7.7.4   Reuse of textile effluent after microfiltration and ion exchange

In textile wet-processing, different types of machinery are used for different types of processes. A thorough study is carried out to characterize the effluents generated from different machines for each process steps in terms of pH, color, COD, SS, hardness, and conductivity. Then the different streams are defined whether it is usable or non-usable to decide for reusability. The layout of the machinery is also to be considered for reuse of water after recycling at the shortest distance. Here, the main aim of the study was to use simple methods like microfiltration, ion exchange, and airing to reduce the COD if necessary to recycle the wastewater at a low cost and to reuse in nearby machines. Reuse of such wastewater results in water-saving, energy-saving, and lowering the cost of effluent treatment [8].

7.7.5   Reuse of wastewater by solar-based photo-Fenton treatments

Solar radiant energy is abundant in tropical countries like India in most of the year. A study conducted on solar energy where evaporation ponds can be used to expose high TDS 5000–10,000 mg L^{-1} effluents for complete evaporation and found to be the most economical way for a zero liquid discharge (ZLD) [34].

Modern techniques like reverse osmosis filtration, microfiltration, ultrafiltration, and nanofiltration are expensive and technologically sophisticated processes. Researchers are always looking for a combination of operations which can be used to treat wastewater effectively with ease. A study based on solar energy assisted photo-Fenton on exhaust dyebaths was found economical and useful where a trial has taken on both laboratory scale (250 ml) and pilot-scale (4 L). Dyebaths were collected from acrylic, wool, polyamide dyeing, polyester dyeing, and reactive dyeing and combined to treat with this process. At a pH of 2.7 adjusted by sulfuric acid, 10 mg L^{-1} iron (II) sulphate was added to the effluent. After homogenization, H_2O_2 was added at a concentration of 60–900 mg L^{-1}. The photo-Fenton reaction was continued in sunlight till the H_2O_2 completely decomposes. Then NaOH was used to adjust pH 8–8.5 for precipitation of $Fe(OH)_3$ which further filtered out. The process removed color in the range of 64–98% and mineralized 60–80% organic matter. After reuse of the recycled water in dyeing, the shades produced were comparable with standard process shades [31].

7.7.6   Direct reuse of textile wastewater in scouring-bleaching

For knitted fabric processing, mostly batch process is followed where it consumes a lot of water and generates too much effluent. The process workflow is varied from industry to industry and fabric to fabric. A general process sequence for knitted fabric processing from greige fabric loading to completion of dyeing stepwise is mentioned as follows: (1) demineralization: a process of removal of metal ions to avoid catalytic damage in the process; (2) scouring and bleaching: a process to improve absorbency and whiteness to facilitate further processes; (3) hot wash; (4) rinsing; (5) neutralization; (6) enzyme treatment: a process to remove short fiber from fabric surface to make it lustrous, smooth, and soft; (7) hot wash; (8) rinsing; (9) dyeing; (10) rinsing; (11) hot wash; (12) soaping; (13) hot wash; (14) hot wash; (15) rinsing; (16) rinsing; and (17) finishing: a process to apply softeners or any finish required. From the above 17 steps effluent collected from steps 1, 4–8, 15, and 16 to use directly in scouring and bleaching of new fabric. The fabric bleached with used water was dyed and compared with the control fabric. It is found that used water from step 4, 5, 7, and 8 can be reused without any treatment. Used water from 1, 6, 15, and 16 are not to be recommended for reuse. To minimize freshwater intake, direct reuse of wastewater from a few streams in the process is possible. This study was found useful in terms of saving water, energy, and effluent treatment cost [32].

7.7.7   Recycling of water, salt, and surfactant from spent acid dyebath

During the dyeing of polyamide fibers with acid dyes, it is found that 90% of dyes consumed and remaining discharged in the effluent. The discharged spent dyebath is a mixture of acid used to adjust pH, sodium sulphate used as retarding agent, and surfactants used for leveling purposes. To reduce the pollution and save water a study was carried out to recycle the spent acid dyebath. A combined process of adsorption and advanced oxidation was found effective in this study. Chitosan adsorption and UV-Fenton advanced oxidation were used for this purpose (CAAOP). The spent acid dyebath was passed through a chitosan column which resulted in complete adsorption of the dye. Permeate from the column was tested for pH, the concentration of salt, and surfactants so that these can be adjusted as per requirement before reuse. Once the column was found saturated, the dyes were desorbed with a dilute sodium hydroxide solution. The desorbed dye solution was treated with UV-Fenton AOP before discharge. This study claimed a saving of water, sodium sulphate, and surfactant in the order of 87.4%, 91.7%, and 50.1%, respectively [21].

7.7.8 Reuse of textile effluent treated with ozone-BAF, UF, and RO

Many advanced treatment processes were studied by the researchers who were practiced in a laboratory scale but found difficult to implement in production. Few of them are electrochemical treatments, catalytic ozonation, ultrasonic-assisted ozone oxidation, photocatalytic oxidation with UV/H_2O_2, Fenton/Photo-Fenton oxidation, etc. A study was carried out on a pilot scale with ozone biological aerated filters (Ozone-BAF) followed by ultrafiltration (UF) and reverse osmosis filtration (RO). The study claimed the dual advantage as RO permeate being suitable for the reuse and the concentrate to be discharged is within the norms. The influent used for this treatment was secondary effluent after coagulation, sedimentation, anaerobic-aerobic activated sludge process with low suspended solid contents. The influent was passed through a two-stage ozone-BAFs. The ozone-BAF was designed with two layers known as the ozonation layer and biodegradation layer with a ratio of 1:5 to 1:8 by volume. The ozonation layer was made up of manganese catalytic grains and ceramic pellets. The manganese ore grain is rich in manganese oxide and acts as an ozone catalyst. The manganese ore grain forms oxygen species or intermediates, to decompose the organic molecules. An ozonation layer used to facilitate further decomposition with biodegradation layer. The RO permeate to concentrate ratio was 3:2. The manufacturer of the system claimed the total investment for a plant with a capacity of 5000 m³ per day, the payback period of 8.6 years with permeate capacity 3000 m³ per day [14].

7.7.9 Reuse of wastewater in reactive dyeing after electrochemical treatment

Reactive dyes are very popular due to their color brilliancy and overall fastness properties. The worldwide consumption is 20–30% of the total market. Fifty percent of the dyes available have azo groups, and similarly, most of the reactive dyes having azo groups (R–N=N–R') followed by anthraquinone. The presence of aromatic rings in the structure not only enhances the chemical stability but also resists to microorganism attack, and hence it is challenging to decompose. Adsorption is the best option for this type of dye removal from the effluent but it requires a further desorption and degradation process. Enzymatic decomposition is quite effective but they need positive control on temperature and pressure to avoid denaturalization. Therefore, the electrochemical treatment method is found a most suitable substitution for removal of dyes where electron used as a clean reagent and there is no residue developed in the process. In this method described by researchers, two electrodes were used called cathode and anode and the rate of decomposition depends on the type of electrode used, surface area, current applied, etc. A study was carried out both in the laboratory (2 L) and pilot-scale (400 L) using such an electrochemical method. The electrodes used for lab having stainless steel cathode and Ti/Pt anode with active surface 60 cm^2 and current density used 177 mA cm^{-2}. For the pilot-scale, the anode was made up of Ti/PtOx with the active surface of 1000 cm^2 and the current density used 70 mA cm^{-2}. The result showed effective color removal and suitability of reuse directly after the treatment with the only objection with the alkaline pH of the solution. When the pH was adjusted to 3.0, it gives an excellent result but practically, it can consume a huge amount of acid. At neutral pH, the result was satisfactory and workable. This process was reported to save 70% of water and 60% of salt from spent dyebath and 100% water with 15% salt from the first washing bath [30].

7.7.10   Recycling of textile wastewater using AOP and its reuse

Many advanced oxidation processes are used by researchers to treat textile wastewater. Ozonation is one of those who basically generates hydroxyl radical (HO*) used as a powerful oxidizing species having the capability to oxidize refractory compounds. After the generation of the radicals, they attack organic compounds (R) by radical addition, hydrogen abstraction, and electrotransfer reactions, as mentioned below in Eqs. (7.1–7.3), respectively.

$$R + HO^* \rightarrow ROH \tag{7.1}$$

$$R + HO^* \rightarrow R^* + H_2O \tag{7.2}$$

$$R^n + HO^* \rightarrow R^{n-1} + OH^- \tag{7.3}$$

A study based on AOP using ozone (O_3), a combination of O_3–ultraviolet light, and O_3–ultraviolet light–H_2O_2 which was applied on effluents from pad batch dyeing showed the efficient removal of color. All three processes removed color above 95%, but when ozonation was combined with UV and both UV–H_2O_2, removal of COD was enhanced. In all cases, reuse of the recycled water resulted in satisfactory dyeing results [11].

7.7.11   Wastewater recycling using ZLD

The conventional bio chemical process removes up to 90% of organic pollutants from wastewater while the removal of COD and TDS is backbreaking. The best alternative to this is RO. A study shows that the incorporation of RO with electrochemical oxidation (EO) and bipolar membrane electrodialysis (BMED) could result in the recycling of 97% of pure water. Maximum concentrates were separated by RO, which were decomposed by electrochemical oxidation. BMED produces aqueous acid and base solution which are reusable both in effluent treatment and wet processing. The discharge of BMED was also used as process water. RO provided 70% permeate and BMED 27% while rest converted to acid and base. The possibility of 100% utilization of effluent was claimed using the reported method [40].

7.7.12   Direct reuse of wastewater in pre-treatment of cotton

To tackle the water scarcity issue in the textile industry, many researchers are continuously working where the aim is to adopt clean technology. Clean technology refers to a reduction in water usage and reuse of water. Cotton textiles are passed through several processes during their wet processing and there is a lot of demand for water as discussed in the beginning. In this context, the direct reuse of wastewater without any treatment within the system will be most beneficial. Direct reuse of wastewater from scouring and bleaching bath in desizing bath was studied. Desizing process can be done by Rot steeping, acid steeping, enzymatic desizing, and oxidative desizing. The wastewater generated from scouring and bleaching contains residual alkali and hydrogen peroxide. The concentration can be checked and adjusted with the addition of required alkali and peroxide to desize a new fabric. The study shows a comparative Tegewa rating (used to test the extent of desizing in Tegewa Scale 1–9), absorbency, whiteness index, and dyeing results in case of regular processed and the fabric processed using direct reuse of wastewater. The reported process showed a saving of 50% water, 19% of chemicals, an overall 40% saving in process cost [13].

7.8   Comparison of research work

For reuse of textile wastewater with different treatments suggested by researchers discussed above are tabulated here in Table 7.5 in which type of effluent used, treatments given, type of fabric used, benefits, and reference of research work is noted. All treatments are aimed at water savings in textile chemical processing.

Table 7.5 Comparison of research works

Sr. no.	Effluent reused	Treatments	Fabric	Benefits	References
1	Exhaust dyebath, end-of-line mill effluent	UF, IE, coagulation, sedimentation	Woolen, PET, and blends	Water and surfactants saving	[36]
2	Composite effluent	Ozonation	Knitted fabric	Water-saving	[25]
3	Exhausted dyebath and subsequent rising	Nanofiltration	Cotton, PET	22% Water saving	[6]
4	Mixed effluent except desizing and dyeing	Microfiltration, IE, Airing	Cotton	Saving of water, energy, and ETP cost	[8]
5	Exhaust dyebath from acid, disperse and reactive dyeing	Solar based photo-Fenton	Cotton, PET, Acrylic	Water-saving	[31]
6	Rising baths from scouring, bleaching, neutralization, and final sage dyeing	No treatment	Knitted fabric	Water saving, direct reuse	[32]
7	Spent acid dyebath	Chitosan adsorption, UV-Fenton AOP	Wool	Saving in water, salt, and surfactant	[21]
8	Composite effluent	Ozone-BAF, UF, RO	–	Water-saving	[14]
9	Reactive dyeing exhaust dyebath	Electrochemical treatment	Cotton	Water-saving	[30]
10	Pad batch dyeing effluent	O_3–UV–H_2O_2	Knitted fabric	Water-saving	[11]
11	Composite effluent	RO-EO-BMED	–	Water-saving	[40]
12	Scouring and bleaching wastewater	No treatment	Cotton	Water and chemical saving	[13]

7.9   Conclusions

Water and energy saving is the most important criteria for any industry; however, for textile industries, water conservation is the main focus area as it also leads to energy conservation. It is essential to develop efficient, cost-effective processes to meet the norms for effluent discharge under pollution control act by local authorities. As discussed in this review, the closed-loop recycle system is helpful to save water and detergent but it is only viable for the large volume of production. A scientific approach to wastewater recovery claimed 22% of water-saving at the cost of the ozonation process. Although nanofiltration is found useful for reuse of used water, it is not suitable for saline water with NaCl. Microfiltration and ion exchange using the zeolite system is helpful for specific wastewater streams. Solar assisted treatment is not a continuous system to operate. Implementation of RO in ZLD is the superior process among all treatments suggested but it may not be affordable for industries of all sizes mainly due to energy intensiveness and high operational cost. As most of the treatments require some treatments, direct reuse of wastewater, and the development of various effluent treatment methods remain the quest for research. Direct reuse of wastewater in the process where ever it is possible is the most economical and measurable approach to reduce effluent load and save water. With a thorough study of the process cycles and effluent characteristics, direct reuse of wastewater in the facility considering the environmental aspects and final product quality will not only reduce the cost of operation but save water leading to a sustainable process.

References

1. Adegoke, K. A.; Bello, O. S. Dye sequestration using agricultural wastes as adsorbents. *Water Resour. Ind.* **2015**, *12*, 8–24. https://doi.org/10.1016/j.wri.2015.09.002.

2. Barbosa, T. R.; Foletto, E. L.; Dotto, G. L.; Jahn, S. L. Preparation of mesoporous geopolymer using metakaolin and rice husk ash as synthesis precursors and its use as potential adsorbent to remove organic dye from aqueous solutions. *Ceram. Int.* **2017**, (August), 0–1. https://doi.org/10.1016/j.ceramint.2017.09.193.

3. Bethi, B.; Sonawane, S. H.; Potoroko, I.; Bhanvase, B. A.; Sonawane, S. S. Novel hybrid system based on hydrodynamic cavitation for treatment of dye waste water: A first report on bench scale study. *J. Environ. Chem. Eng.* **2017**, *5*(2), 1874–1884. https://doi.org/10.1016/j.jece.2017.03.026.

4. Bilal, M.; Asgher, M.; Iqbal, M.; Hu, H.; Zhang, X. Chitosan beads immobilized manganese peroxidase catalytic potential for detoxification and decolorization of textile effluent. *Int. J. Biol. Macromol.* **2016**, *89*, 181–189. https://doi.org/10.1016/j.ijbiomac.2016.04.075.

5. Chen, L.; Wang, L.; Wu, X.; Ding, X. A process-level water conservation and pollution control performance evaluation tool of cleaner production technology in

textile industry. *J. Clean. Prod.* **2017**, *143*, 1137–1143. https://doi.org/10.1016/j.jclepro.2016.12.006.

6. De Vreese, I.; Van der Bruggen, B. Cotton and polyester dyeing using nanofiltered wastewater. *Dyes Pigments.* **2007**, *74*(2), 313–319. https://doi.org/10.1016/j.dyepig.2006.02.014.

7. Dojčinović, B. P.; Roglić, G. M.; Obradović, B. M.; Kuraica, M. M.; Kostić, M. M.; Nešić, J.; Manojlović, D. D. Decolorization of reactive textile dyes using water falling film dielectric barrier discharge. *J. Hazard. Mater.* **2011**, *192*(2), 763–771. https://doi.org/10.1016/j.jhazmat.2011.05.086.

8. Erdumlu, N.; Ozipek, B.; Yilmaz, G.; Topatan, Z. Reuse of effluent water obtained in different textile finishing processes. *Autex Res. J.* **2012**, *12*(1), 23–28. https://doi.org/10.2478/v10304-012-0005-9.

9. Firdhouse, M. J.; Lalitha, P. Nanosilver-decorated nanographene and their adsorption performance in waste water treatment. *Bioresour. Bioprocess.* **2016**, *3*. https://doi.org/10.1186/s40643-016-0089-5.

10. Ghezzar, M. R.; Abdelmalek, F.; Belhadj, M.; Benderdouche, N.; Addou, A. Enhancement of the bleaching and degradation of textile wastewaters by Gliding arc discharge plasma in the presence of TiO_2 catalyst. *J. Hazard. Mater.* **2009**, *164*(2–3), 1266–1274. https://doi.org/10.1016/j.jhazmat.2008.09.060.

11. Guyer, G. T.; Nadeem, K.; Dizge, N. Recycling of pad-batch washing textile wastewater through advanced oxidation processes and its reusability assessment for Turkish textile industry. *J. Clean. Prod.* **2016**, *139*, 488–494. https://doi.org/10.1016/j.jclepro.2016.08.009.

12. Hangargekar, P. A.; Takpere, K. P. A case study on waste water treatment plant, CETP (common effluent treatment plant). *IJIRAE.* **2015**, *2*(11), 34–39.

13. Harane, R.; Adivarekar, R. A frugal way of reusing wastewater in textile pre-treatment process. *J. Water Process Eng.* **2017**, *16*, 163–169. https://doi.org/10.1016/j.jwpe.2017.01.002.

14. He, Y.; Wang, X.; Xu, J.; Yan, J.; Ge, Q.; Gu, X.; Jian, L. Application of integrated ozone biological aerated filters and membrane filtration in water reuse of textile effluents. *Bioresour. Technol.* **2013**, *133*, 150–157. https://doi.org/10.1016/j.biortech.2013.01.074.

15. Hong, G.-B.; Wang, Y.-K. Synthesis of low-cost adsorbent from rice bran for the removal of reactive dye based on the response surface methodology. *Appl. Surf. Sci.* **2017**, *423*, 800–809. https://doi.org/10.1016/j.apsusc.2017.06.264.

16. IBEF. Textile Industry & Market Growth in India; **2017**. Retrieved August 1, 2017, from https://www.ibef.org/industry/textiles.aspx.

17. Janveja, B.; Kant, K.; Sharma, J. A study of activated rice husk charcoal as an adsorbent of Congo red dye present in textile industrial waste. *Jpafmat.* **2008**, *8*(1), 12–15.

18. Kamil, A.; Abdalrazak, F.; Halbus, A. Adsorption of bismarck brown R dye onto multiwall carbon nanotubes. *J. Environ. Anal. Chem.* **2014**, *1*(1), 1–6. https://doi.org/10.4172/JREAC.1000104.

19. Karadagi, B. S.; Hiremath, G. M. Removal of COD and color from textile waste water using sawdust as an adsorbent. *Int. J. Innov. Res. Sci. Eng. Technol.* **2016**, *5*(6), 9593–9597.

20. Khandegar, V.; Saroha, A. K. Electrocoagulation for the treatment of textile industry effluent: A review. *J. Environ. Manag.* **2013**, *128*, 949–963. https://doi.org/10.1016/j.jenvman.2013.06.043.

21. Li, C. H.; He, J. X. Advanced treatment of spent acid dyebath and reuse of water, salt and surfactant therein. *J. Clean. Prod.* **2013**, *59*, 86–92. https://doi.org/10.1016/j.jclepro.2013.06.049.

22. Mane, V. S.; Babu, P. V. V. Studies on the adsorption of Brilliant Green dye from aqueous solution onto low-cost NaOH treated saw dust. *Desalination.* **2011**, *273*(2–3), 321–329. https://doi.org/10.1016/j.desal.2011.01.049.

23. Mane, V. S.; Babu, P. V. V. Kinetic and equilibrium studies on the removal of Congo red from aqueous solution using Eucalyptus wood (*Eucalyptus globulus*) saw dust. *J. Taiwan Inst. Chem. Eng.* **2013**, *44*(1), 81–88. https://doi.org/10.1016/j.jtice.2012.09.013.

24. Ministry of Water Resources. Water the resource – General facts; **2014**. Retrieved August 1, 2017, from http://wrmin.nic.in/forms/list.aspx?lid=297.

25. Orhon, D.; Germirli Babuna, F.; Kabdaşlí, I.; Insel, F. G.; Karahan, Ö.; Dulkadiroğlu, H.; … Yediler, A. A scientific approach to wastewater recovery and reuse in the textile industry. *Water Sci. Technol.* **2001**, *43*(11), 223–231.

26. Patel, H.; Vashi, R. T. In *Characterization and Treatment of Textile Wastewater* (1st Edition); **2015**. 225 Wyman Street, Waltham, MA 02451, USA: Elsevier publications.

27. Peláez-Cid, A. -A.; Herrera-González, A.-M.; Salazar-Villanueva, M.; Bautista-Hernández, A. Elimination of textile dyes using activated carbons prepared from vegetable residues and their characterization. *J. Environ. Manag.* **2016**, *181*, 269–278. https://doi.org/10.1016/j.jenvman.2016.06.026.

28. Poornima Parvathi, V.; Umadevi, M.; & Bhaviya Raj, R. Improved waste water treatment by bio-synthesized graphene sand composite. *J. Environ. Manag.* **2015**, *162*, 299–305. https://doi.org/10.1016/j.jenvman.2015.07.055.

29. Robertson, A.; Nghiem, L. D. Treatment of high TDS liquid waste : Is zero liquid discharge feasible ? *J. Water Sustain.* **2011**, *1*, 1–11.

30. Sala, M.; Gutiérrez-Bouzán, M. C. Electrochemical treatment of industrial wastewater and effluent reuse at laboratory and semi-industrial scale. *J. Clean. Prod.* **2014**, *65*, 458–464. https://doi.org/10.1016/j.jclepro.2013.08.006.

31. Sanz, J. F.; Monllor, P.; Vicente, R.; Amat, A. M.; Arques, A.; Bonet, M. Exploring reuse of industrial wastewater from exhaust dyebaths by solar-based photo-Fenton treatment. *Text. Res. J.* **2013**, *83*, 1327–1334. https://doi.org/10.1177/0040517512467061.

32. Shaid, A.; Osman, S.; Hannan, A.; Bhuiyan, M. A. R. Direct reusing of textile wastewater in scouring-bleaching of cotton goods devoid of any treatment. *Int. J. Eng. Res. Dev.* **2013**, *5*(8), 45–54.

33. Shalaby, N. H.; Ewais, E. M. M.; Elsaadany, R. M.; Ahmed, A. Rice husk templated water treatment sludge as low cost dye and metal adsorbent. *Egypt. J. Petrol.* **2017**, *26*(3), 661–668. https://doi.org/10.1016/j.ejpe.2016.10.006.

34. Sivakumar, K. K.; & Dheenadayalan, M. S. Studies on the utilization of solar energy in the effluent treatment. *Eur. Chem. Bull.* **2012**, *1*(5), 146–149. https://doi.org/10.17628/ECB.2012.1.146.

35. Tong, T.; Elimelech, M. The global rise of zero liquid discharge for wastewater management: Drivers, technologies, and future directions. *Environ. Sci. Technol.* **2016**, *50*(13), 6846–6855. https://doi.org/10.1021/acs.est.6b01000.

36. Turnbull, R. H.; Groves, G. R.; & Buckley, C. A. Closed looped recycle systems for textile effluents. *Water Environ. Feder.* **1979**, *51*(3), 499–517.

37. Vergili, I., Kaya, Y., Sen, U., Gonder, Z. B., & Aydiner, C. Techno-economic analysis of textile dye bath wastewater treatment by integrated membrane processes under the zero liquid discharge approach. *Resour. Conserv. Recycl.* **2012** *58*, 25–35. https://doi.org/10.1016/j.resconrec.2011.10.005

38. Vuono, D.; Catizzone, E.; Aloise, A.; Policicchio, A.; Agostino, R. G.; Migliori, M.; Giordano, G. Modelling of adsorption of textile dyes over multiwalled carbon nanotubes: Equilibrium and kinetic. *Chin. J. Chem. Eng.* **2017**, *25*(4), 523–532. https://doi.org/10.1016/j.cjche.2016.10.021.

39. Wawrzkiewicz, M.; Bartczak, P.; Jesionowski, T. Enhanced removal of hazardous dye form aqueous solutions and real textile wastewater using bifunctional chitin/lignin biosorbent. *Int. J. Biol. Macromol.* **2017**, *99*, 754–764. https://doi.org/10.1016/j.ijbiomac.2017.03.023.

40. Yao, J.; Wen, D.; Shen, J.; Wang, J. Zero discharge process for dyeing wastewater treatment. *J. Water Process Eng.* **2016**, *11*, 98–103. https://doi.org/10.1016/j.jwpe.2016.03.012.

8

Sustainable approaches towards decontamination of industrial effluents: Recent developments in adsorptive removal of malachite green dye from aqueous solutions

M. Vasudevan[a,*] **and N. Natarajan**[b]

aDepartment of Civil Engineering, Bannari Amman Institute of Technology, Sathyamangalam, Erode-638401, Tamil Nadu, India

E-mail: devamv@gmail.com

bDepartment of Civil Engineering, Dr. Mahalingam College of Engineering and Technology, Pollachi-642003, Tamil Nadu, India.

E-mail: itsrajan2002@yahoo.co.in. Phone: 04259-236030; Fax: 04259-236060

*Corresponding Author: Dr. M Vasudevan

Abstract: The widespread application of combined techniques for pre-treatment of industrial effluents has invoked the researchers to identify economic and sustainable approaches for improving the sorption capacity and regeneration potential of adsorbents. The conversion of wastes into materials that can potentially improve the environment would be advantageous in response to ecological constraints for the sustainable effluent treatment strategies followed by the industries using dyes. This study presents a roadmap towards achieving sustainable decontamination of textile effluents by adsorptive removal, inferred from various natural and synthetic engineered materials of promise. It is essential to compare their characteristics and performance across laboratory and field conditions in order to pursue realistic strategies towards sustainability.

Keywords: low cost adsorbent; malachite green dye; activated carbon; textile effluent; agricultural waste; industrial waste.

8.1   Introduction

Industries such as textiles, paper, leather, plastics, tannery, etc. utilize dyes to color their products and also consume considerable amount of water. Consequently, they generate a large amount of colored wastewater [11]. On an average, 40 L of water is required to produce 1 kg of textile product, leading to pose large environmental footprint [74]. The textile industry being one of the largest industrial consumers of water is also the greatest generator of hazardous liquid effluent as a result of highly energy–water–chemical intensive processes [17]. The global textile and apparel industry (including textile, clothing, footwear, luxury, and fashion items) is currently worth nearly $3,000 trillion [71]. The presence of dyes in the effluent is a major cause of concern for environmental livelihood due to its multi-faceted adverse effects. It is highly heterogeneous as the nature of disposed chemicals can vary from different finishing processes. It is estimated that more than ten thousand commercially available dyes with more than 70,000 tonnes of dyestuff produced annually [12–14]. On an average, about 15% of the global production of dyes is observed to be escaping to the environment during the dyeing process as toxic effluent [10,18].

Malachite green (MG) is a cationic dye widely used as coloring agent in textile, apparel, tannery, distillery, and paper industries; therapeutic agent in aquaculture, fungicide, parasiticide, and disinfectant in fishery industries; additive in food industries and also as a medical disinfectant in pharmaceutical industries [20–22]. Chemically, it is triarylmethane dye with formula $C_{23}H_{25}ClN_2$ (molecular weight of 364.9 g mol^{-1}), having a strong absorption band of cation at 621 nm resulting in intense green color. Its occurrence in the natural streams can cause serious harm to aquatic life by increasing chemical oxygen demand, thereby hindering the photosynthetic phenomena through reduction of light penetration [15,16]. Consumption of MG has many adverse effects as it possesses carcinogenic, genotoxic, mutagenic, and teratogenic properties [8,9,19]. It is a Class II health hazard with oral LC50 of 80 mg kg^{-1} when tested on mouse. Therefore, the treatment of the effluent containing such a dye is of great interest to protect the environmental flow of the receiving water body as well as health and safety of downstream consumers.

Many conventional treatment methods have been adopted for the removal of dyes from wastewater during the past few decades such as cation exchange, electrochemical degradation, biological treatment, chemical oxidation, photocatalytic degradation, membrane separation, etc. However, conventional wastewater treatment processes cannot remove the dyes and pigments from water because of their complex molecular structure, fair stability to light and heat, and recalcitrant nature [13]. Forgacs et al. [60] had given an extensive but generic review on selectivity of synthetic dye removal systems such as adsorption, chemical oxidation, photo-degradation, and microbial decoloration. Either being inefficient or expensive on a sustainable point of view, most of these processes are not well adopted in large scale [7].

Among economical and eco-friendly treatment technologies, adsorption has rapidly gained prominence as a sustainable choice for treating industrial and domestic effluent [26,27,61]. It has been considered superior to other methods in terms of cost, flexibility, simplicity of design, ease of operation, material recovery, and insensitivity to toxic pollutants [1,6]. Adsorption is a well-known equilibrium separation process and is being employed widely for large scale biochemical, chemical, environmental recovery, and purification applications [2–5].

Extensive research has been carried out during the last 20 years to find low cost, high capacity adsorbents for the removal of organic dyes from textile effluents. It includes use of natural materials (bentonite, clays, goethite, and kaolin), preparation of biosorbents, and activated carbons from agricultural products (fruits, seeds, and plant parts) and industrial waste materials (bottom ash, fly ash, sludge, and other byproducts). Although commercially available activated carbon is well accepted for its excellent surface properties for removing both organic as well as inorganic pollutants from wastewater, alternative methods of preparations have to be adopted for reducing initial and regeneration costs.

The aim of this present review article is to provide a state-of-the-art summary of the studies during the past two decades dealing the usage of low-cost natural, agricultural, and industrial waste materials as adsorbents for removal of MG dyes from the aqueous solution. The study is intended to update the researchers about identification of new materials, advancement in activation methodology, development of surface characterization techniques, selection of appropriate mass transfer models, and latest insights into the mechanisms of adsorption. Based on this, this study presents an overview of the research studies centered on the dye, malachite green (Fig. 8.1).

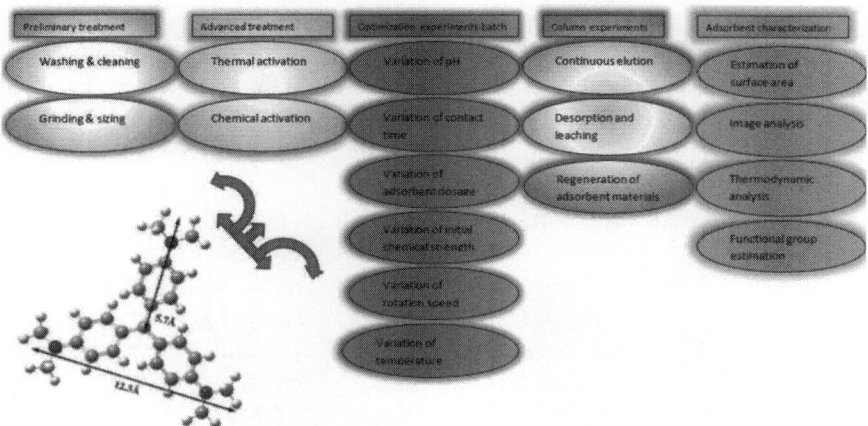

Figure 8.1   Overview of research studies on malachite green dye

8.2   Identification of non-conventional low-cost adsorbents

A number of agricultural products/wastes and industrial waste products have been proposed by many researchers for the removal of MG from aqueous solutions. Quite often, the 'successful' discovery of an adsorbent is acclaimed on the base of its adsorption capacity or maximum removal efficiency for a particular pollutant. However, it seems insufficient on a sustainable point of view keeping in view of the energy and cost involved in pre-treatment steps, method of activation (if any) and ease of material recovery (regeneration) potential. With this focus, a comparison has been made about the recent developments in non-conventional low cost adsorbents for MG removal.

8.2.1   Materials from agricultural solid wastes

Agricultural materials are considered to be low-cost since they are generally abundant in their respective natives, can be easily collected, require less processing and generally do not create harmful by-products. Many solid agricultural wastes such as shells, fibers, fruit peels, seeds, saw dust, etc. have been widely used as adsorbents for removal of organic pollutants. Details of the pre-treatment steps, activation methods, optimum conditions for batch experiments, and observed maximum adsorption capacity have been provided in Table 8.1.

8.2.1.1   *Parts from fruits and grains*

Many valuable materials from plants and fruits such as peels, bran, husks, and shells were widely tried by many researchers in the past two decades. Papinutti et al. [56] studied adsorption of MG onto wheat bran and degradation of white rot fungi using MG as substrate. There was no pre-treatment or activation steps for wheat bran except drying at 60 °C for 24 h. Very similar to the above case, the maximum adsorption capacity was found to be 24 mg g^{-1} at a pH of 5. The concentration of MG varied from 12 mg g^{-1} to 348 mg g^{-1} while the optimum dosage was found to be 1.0 g/10 ml for a contact time of 40 min. The rate of adsorption has significant importance in the real-time application of these systems where costs of operation and removal efficiency are of prime concern. However, it is noted that the time required to attain equilibrium is considerably less in this study compared to many similar batch experiments. They have stated that selection of optimum pH for adsorption and degradation is the critical step for successful remediation strategy.

The adsorption capacity of wheat bran and rice bran were also investigated by Wang et al. [55] in order to optimize the experimental conditions for achieving maximum removal efficiency. They have reported higher adsorption capacity (68.97 mg g^{-1} for rice bran and 66.57 mg g^{-1} for wheat bran) under

Table 8.1 Batch and continuous/extended experiments

Adsorbent	Experimental conditions					Isotherm model	Isotherm parameters	Kinetic model	Kinetic parameters	Thermodynamic parameters	Overall removal efficiency	Max. adsorption capacity (mg g⁻¹)	Remarks	Reference
	pH	Dosage (g L⁻¹)	Time (min)	Temperature (°C)	IC of MG									
Industrial waste materials														
Blast furnace slag	6–10	10	360–480	30–50	1×10^{-4} M to 1×10^{-3} M	L, F	Qo=10.52e5; b=2.13e-3; Kf=19.95e-6; n=2.1	FO	7.48e-3 min^{-1}	$-\Delta G$=30.90 kJ mol^{-1} $-\Delta H$=18.70 kJ mol^{-1} $-\Delta S$=163.0 J mol^{-1} K^{-1}	~100%	0.22	Regeneration in column exp. 93% within five cycles	Gupta et al. [77]
Power plant-generated bottom ash	5.0	4.0	60–120	30	1×10^{-5} to 1×10^{-4} M	L F	Qo=1.618×10^{-3} g^{-1}; b=14.95 L mol^{-1} Kf=3.168×10^{-4} g^{-1}; n=0.1256	FO	Kd=2.303 × 10^{-2} min^{-1}	$-\Delta G$=6.81 $-\Delta H$=24.10 $-\Delta S$=97.71	96%	0.71	Regeneration in column exp. 94% within five cycles	Gupta et al. [20]
Akash kinari coal	7.2	20	110	20–40	5–20 mg L^{-1}	L F	Qo=2.33 mg g^{-1}; b=2.189, RL=0.0028 Kf=0.053, n=0.9164	FO	Kd=0.045 min^{-1}	$-\Delta G$=7.428 $-\Delta H$=6.397 $-\Delta S$=47.16	79–89%	0.8		Khan et al. [69]
Bagasse fly ash	7–11	1.0	240	30	10–250 mg L^{-1}	L, F, RP	kf= 25.149; 1/n=0.423 Kl=0.148 L mg^{-1}; qm=170.33 mg g^{-1}; Rl=0.119 RP-> Kr=2279.2 L g^{-1}; aR=89.151; B=0.583	PSO, IPD	k=0.0168 min^{-1}	$-\Delta G$=33.62	93	170.33 mg g^{-1}		Mall et al. [72]
Fly ash	7.0–7.5	20	80	30	5–20 mg L^{-1}	L F	Qo=1.3717, B=0,0407, RL=0.0306 Kf=0.0797, 1/n=0.6777	FO	4.42e2 min^{-1}	$-\Delta G$=8.07 $-\Delta H$=8.8 $-\Delta S$=2.4	64–87%	0.644 mg g^{-1}	Acid-base dissociation	Khan et al. [69]

Adsorbent	Experimental conditions					Isotherm model	Isotherm parameters	Kinetic model	Kinetic parameters	Thermodynamic parameters	Overall removal efficiency	Max. adsorption capacity (mg g^{-1})	Remarks	Reference
	pH	Dosage (g L^{-1})	Time (min)	Temperature (°C)	IC of MG									
De-inking paper sludge and virgin pulp mill	9.2–10.3	2.5	120	25	10–4000 mg L^{-1}	L, F	Q=982 and 435 KL=0.10 and 0.23 Kf=58.3 and 35.8 1/m= 0.43 and 0.34				98% and 80%	982 mg g^{-1} and 435 mg g^{-1}	Thermal decomposition of MG	Mendez et al. [63]
Fly ash	7	10	40	30–45	10–100 mg L^{-1}	L, F, DR	qm=17.8 mg g^{-1}; KL=0.8146; Kf=7.2545; 1/n=0.6323	PSO	K2=8.98e-2 g mg^{-1} min^{-1}	$-\Delta G$=14.27 $-\Delta H$=84.73 $-\Delta S$=310.9	99.07	4.94 mg g^{-1}		Chowdhury and Saha [67]
Activated sintering- red mud	>3	5	180	25	10–500 mg L^{-1}	L	Qo=336.4 mg g^{-1}, b=0.077 L mg^{-1}	PSO, IPD	1.23e-3 g mg^{-1} min^{-1}	$-\Delta G$=11.76 ΔH=167.8 ΔS=603.2	98.5	93.2 mg g^{-1}		Zhang et al. [78]

similar experimental conditions within an equilibration time of 60–90 min (Table 8.1) without any appreciable treatment (other than oven drying overnight). It is to be noted that adsorption capacity increased with increase in pH as well as adsorbent dosage, while decreased with increase in ionic strength of the solution.

It is observed that silica present on the outer surface of rice husks (of silicon–cellulose membrane) is responsible for insufficient binding between accessible functional groups and various adsorbate ions/molecules, while the inner surface is smooth and may contain wax and natural fats which also affects the adsorption properties of rice husk chemically and physically. Therefore, in their study, Chowdhury et al. [59] used NaOH-treated rice husk in order to improve adsorption properties by removing silica and other impurities from the inner surface of rice husk. The maximum adsorption capacity was 17.98 mg g^{-1} at 40 °C under an optimum pH of 7 for 2 h while varying the initial concentration from 10 mg L^{-1} to 100 mg L^{-1}.

Other interesting materials recovered from agricultural products are the peels for adsorptive removal of MG. Kumar [31] attempted to evaluate the feasibility of removal of MG using lemon peel while Kumar and Porkodi [46] studied about the batch adsorber design using orange peel as adsorbent. Under similar conditions of experiments (Table 8.1), orange peels required a less dosage of 0.009 g/30 ml compared to 0.05 g/30 ml for lemon peel, whereas the adsorption capacity obtained by orange peel is much higher (483.63 mg g) as compared to that of lemon peel (51.73 mg g). It is interesting to note that the adsorbent required no activation steps other than cutting, drying, and grinding.

Santhi and Manonmani [54] prepared activated carbon from *Cucumis sativa* fruit peels for efficient removal of MG from simulated dye wastewater with a removal efficiency of 99.86%. However, considering the energy expenses for preparing the activated carbon as well as dosage required (1 g/50 ml), observed maximum adsorption capacity was much less (36.23 mg g^{-1}) as compared to use of natural orange peels [46].

In a similar attempt, Sharma et al. [41] investigated adsorption of MG using chemically modified potato peels. The maximum adsorption capacity was observed to be 111 mg g^{-1}, which is high enough compared to many untreated adsorbents. After initial washing, drying, and grinding, the peels were dipped overnight in 6% formaldehyde solution, and after drying, subjected to 0.2 N sulfuric acid at a ratio of 1:1 on weight/volume basis. The adsorbent was found to be effective at a pH of 12 and temperature of 27 °C with a dosage of 0.25 g/100 ml.

Chemically modified breadnut peel was effectively used for removal of MG by Chieng et al. [48] with the highest adsorption capacity of 353 mg g^{-1}. After primary drying and blending, the adsorbent was treated with 4.0 M sodium hydroxide at a ratio of 1 g/100 ml for a period of 2 h. The optimum pH was observed to be at 3.6 indicating that adsorption capacity was highly influenced by the electrostatic attraction between the functional groups on the adsorbent surface and cationic dye. It is widely accepted that the degree of adsorption of MG is highly influenced by the medium pH and the surface charge of the adsorbate.

Mittal et al. [32] explored the possibility of using de-oiled soya, a waste product from soya oil extraction mills for removal of MG dye. The adsorbent was prepared with least treatment steps apart from washing and drying. To oxidize the adhering organic material, it was treated with hydrogen peroxide for 24 h and further dried to remove moisture. They have observed maximum removal efficiency of 100% at a pH of 5 for 1×10^{-5} M to 1×10^{-4} M MG solution with a dosage of 0.05 g/25 ml within the contact time of 24 h. The maximum adsorption capacity was found to be 20.44 mg g^{-1} under the optimum conditions. Sonawane and Shrivastava [52] observed that adsorptive capacity of maize cob powder was quite comparable (80.64 mg g^{-1}) with that of any agricultural product having low dosage requirement (0.5–12 g L^{-1}) under the normal environmental conditions (at pH 8 and 27 °C for 80 min) without any pre-treatment. For these materials, the uptake of dye was found to increase with increasing contact time and initial dye concentration and to decrease with an increasing adsorbent dosage.

Ahmad and Kumar [23] studied the adsorption of MG from aqueous solution onto treated ginger waste using batch and column methods. In each batch experiment, 50 ml dye solution of known concentration (5–20 mg L^{-1}) and known amount of (0.05 g) of treated ginger waste was taken at the constant temperature. The uptake of MG increased with increase in contact time as well as increase in initial dye concentration. MG dye removal was optimum when pH was in the alkaline range. Fixed bed column studies indicated the maximum breakthrough capacity and exhaustive capacity for 0.2 g, 0.3 g, 0.5 g of treated ginger waste to be 7.5 mg, 11.9 mg, 18.1 mg and 10.1 mg, 15.9 mg, 24.3 mg, respectively.

Baek et al. [24] studied the adsorption of MG on degreased coffee beans using batch experiments with a dosage of 0.1 g/5 ml, pH of 10–12 and a contact time of 4 h. The adsorption capacity at equilibrium was 23 mg g^{-1} while the percentage of adsorption decreased from 84.6% to 78.6% as the initial MG concentration increased from 25 mg L^{-1} to 100 mg L^{-1}. Degreased coffee beans were found to have more adsorption capacity compared to ordinary coffee beans. But desorption of the degreased coffee beans using 0.1 M HCl was not very effective. Considering overall expenses and efficiency, degreased coffee beans are cheaper, easily available, and need minimal pre-treatment compared to activated carbon and other natural materials.

Mango seed husks were studied for removal of MG under high initial dye concentrations (100–500 mg L^{-1}) in batch experiments [53]. The optimum dosage was 0.1 g/20 ml under an acidic pH of 5 and room temperature (30 °C) over a period of 2 h, resulting in adsorption capacity of 47.9 mg g^{-1}. Since there is no pre-treatment required, the observed performance of the adsorbent seems to be satisfactory. Jalil et al. [29] demonstrated that bivalve shell (BS) and *Zea mays* L. husk leaf (ZHL) can be used as an effective adsorbent for the removal of MG from aqueous solution. Pre-treatment of the raw ZHL with BS effectively enhanced its electro-negativity, which led to an increased adsorption capacity (81.5 mg g^{-1}) and removal efficiency (98%) after 30 min of contact time under the dosage of 0.01 g/200 ml at pH of 6. The lower dosage requirement, short period of contact time, and wide range of initial dye concentration (10–200 mg L^{-1}) certainly marks its candidature for a sustainable alternative.

In addition to this, many other fruit shells were introduced to examine their adsorptive capacity for MG from aqueous solution. Saha et al. [33] reported maximum adsorption capacity of 1.951 mg g^{-1} with dried tamarind fruit shell at a pH of 5 and temperature of 30 °C. A dosage of 1 g/100 ml of adsorbent was sufficient for a period of 60 min to remove various initial concentrations of MG (10–200 mg L^{-1}) with a maximum removal efficiency of 98.8%. However, since the adsorptive capacity is very low, the regeneration of adsorbent is not economically feasible.

Similarly, Ozdes et al. [49] identified almond shell as a potential adsorbent at a dosage of 0.1 g/10 ml for MG varying from 70 mg L^{-1} to 1150 mg L^{-1}. The maximum adsorption capacity was observed to be 29 mg g^{-1} at a pH of 4.5 and temperature of 40 °C. As the waste material can be easily collected and used for adsorption without any pre-treatment, this also can be accepted to be a viable option. Dahri et al. [25] showed that walnut shell has the potential to be regenerated and reused while maintaining its ability to adsorb more than 50% MG even after five cycles when washed with 0.1 M NaOH. The optimized condition for adsorption process in this study was carried out at a dosage 0.03 g/20 ml dye, at room temperature, ambient pH, and agitation rate of 250 rpm for 2 h. The highest adsorption capacity (90.8 mg g^{-1}) and percentage removal (87%) are attributed to its greater surface area and reduced particle size (<355 mm).

Wood apple shell, a fruit-food solid waste, was successfully utilized as a low cost alternative adsorbent for the removal of MG [34]. The removal of MG dye was found to be 98.87% with initial concentration 100 mg L^{-1} at pH 7–9 in about 210 min while the maximum adsorption capacity (80.645 mg g^{-1}) was comparatively good. Considering the ambient environmental conditions are conducive for optimum reaction and that no pre-treatment was recommended by the authors, this is also considered as a feasible option.

Kooh et al. [70] studied the adsorption characteristics of jackfruit seed as an adsorbent for the removal of MG dye from the aqueous solution via batch adsorption experiment. Several parameters like contact time, pH, dye concentration, ionic strength, and temperature were investigated. Freundlich model was found to best fit the experimental data, and the maximum mono-layer capacity was determined as 66 mg g^{-1}. The kinetics mechanism followed pseudo-second-order model and the intraparticle diffusion was found not to be the rate-determining step. The authors also used machine learning techniques, namely artificial neural networks and random forest to predict the adsorption process.

Mohammad et al. [74] analyzed the effectiveness of using physic seed hull (*Jatropha curcas* L.) as an alternative low-cost adsorbent for the removal of MG from simulated wastewater. The adsorption behavior such as kinetics, dynamics, and isotherms of the hull was studied. The kinetic data fitted well with the pseudo-second-order kinetic model for the MG adsorption. Langmuir isotherm was found to be the best fit model to describe the adsorption process.

8.2.1.2   Types of saw dust

Saw dust is a waste by-product of the timber industry that is either used as cooking fuel or a packing material. Since they are naturally available in the dry state and can be easily disintegrated to particle form by mechanical crushing, most of the studies have not recommended any pre-treatment for adsorbent preparation.

Garg et al. [26] investigated the adsorption of MG by rosewood saw dust pre-treated with formaldehyde and sulfuric acid. The adsorption efficiency of sulfuric acid treated sawdust (74.5 mg g^{-1}) was observed to be higher than formaldehyde treated sawdust (26.9 mg g^{-1}). Increase in initial dye concentration decreased the percent adsorption by sawdust and remained constant after equilibrium time. For an initial dye concentration of 250 mg L^{-1} and adsorbent dosage of (0.2–1.0 g/100 ml), the removal efficiency increased from 59.6% to 99.8% for the sulfuric acid treated sawdust carbon dose, while it increased from 18.6% to 86.9% for formaldehyde treated saw dust with same adsorbent doses.

Sawdust of *Prosopis cineraria*, a commonly grown perennial tree in dry and arid regions of Northern and Western India, is a waste by-product of timber industry that is either used as domestic fuel or a packing material. Garg et al. [62] observed that adsorption capacity decreased from 65.8 mg g^{-1} to 25.0 mg g^{-1} as the adsorbent dose was increased from 0.2 to 1.0 g/100 ml in the test solution. Based on the flexibility in operational conditions (pH 6–10, temperature 27 °C, contact time 180 min, dosage 4 g L^{-1} and initial concentration from 0 mg L^{-1} to 500 mg L^{-1}), it deserves to be a good alternative biosorbent material. Kumar and Sivanesan [51] obtained adsorption capacity of 36.45 mg g^{-1} of MG (initial concentration of 100 mg L^{-1}) onto rubber wood saw dust, which is comparable with other low cost adsorbents.

Hameed and Khaiary [27] studied the adsorption of MG on rattan sawdust at 30 °C at a dosage of 0.3 g/200 ml for MG initially varying from 25 mg L^{-1} to 300 mg L^{-1} for a contact time of 210 min. The maximum adsorption capacity was reported as 62.71 mg g^{-1} at 30 °C under an acidic pH of 4. Khattari and Singh [30] attempted to use need saw dust for adsorptive removal of MG. The adsorption capacity was found to be 4.35 mg g^{-1} which was quite low compared to many other low cost adsorbents. Even with a low range of dye (6–12 mg L^{-1}) over a contact time of just 14 min considering the favorable conditions of pH 7.2, temperature 25 °C, and dosage of 0.25 g/50 ml, it is substantiated that need saw dust has good potential for adsorption and may be enhanced with suitable thermo-chemical activation methods. Use of beech saw dust as biosorbent for MG was investigated by Witek-Krowiak [44]. Without any treatment, they could observe high adsorption rate (83.21 mg g^{-1}) for a dosage of 0.25 g/100 ml at pH 5 and 20 °C for a period of 120 min. This study was carried for a high range of MG concentration (from 10 mg L^{-1} to 1000 mg L^{-1}).

Adsorption of MG onto camphora saw dust was significantly modified by treating chemically with oxalic acid (280.3 mg g^{-1}), citric acid (22.8 mg g^{-1}), and tartaric acid (157.5 mg g^{-1}) [55]. The pre-treatment consists of washing, drying at 60 °C for 48 h and treating with the chemical in 1:1 (volumetric proportion) for 3 h. The optimum adsorption was achieved at an adsorbent dosage of 0.05 g/50 ml at pH 9 and 60 °C for a period of 720 min. They had observed that the extent of MG adsorption onto modified sawdust increased with increasing organic acid concentrations, pH, contact time, and temperature but decreased with increasing adsorbent dosage and ionic strength.

8.2.1.3   Other plant parts

As a viable alternative, biosorbents are becoming popular due to their low cost, good performance, and environmental benignity. Apart from fruits, use of tree leaves and barks were also used as potential adsorbents for MG removal from aqueous solution. Thermo-chemically modified rice straw has been suggested as a potential adsorbent for MG removal (93%) at pH 4 with a dosage of 2.0 g L^{-1} with a very high adsorption capacity of 256.41 mg g^{-1} at optimum conditions [47]. This is about three times higher than the adsorption capacity of native form of rice straw. Singh and Kaur [50] observed maximum adsorption capacity of rice straw (92 mg g^{-1}) under normal environmental conditions (pH 8, temperature 35 °C, initial concentration of 50 mg L^{-1} and adsorbent dosage 1% w/w). It was observed that by using 0.1 M sodium hydroxide, higher amount of dye (almost 99%) could be regenerated from the adsorbed rice straw surface after three cycles, thereby making it a vital sustainable product (Table 8.1).

Hameed and Khaiary [28] studied the adsorption of MG using oil palm trunk fibers. Oil palm trunk is one of the major solid waste materials generated from palm-oil upstream industry which is rich in lingo-cellulosic compounds. Maximum adsorption (149.35 mg g^{-1}) was obtained at a pH of 8 with a dosage of 0.3 g/200 ml for 120 min at 30 °C. In a similar attempt, dead leaves of plane tree (*Platanus vulgaris*) were investigated as a novel biosorbent of dyes taking aqueous MG solutions as a model system [36]. MG concentration was varied from 5 mg L^{-1} to 50 mg L^{-1}, at a dosage of 0.1–0.25 g/100 ml with pH of 5.5 at **40** °C for a contact time of one day. Without any single pre-treatment steps, they have reported highest adsorption capacity of 89.43 mg g^{-1}, which is quite comparable with many other bio-sorbents.

Use of wood fiber from phoenix tree (*Firmiana simplex*), a deciduous tree in China was attempted by Pan and Zhang [37] without any pre-treatment. It was observed to be effective (adsorption capacity of 142.4 mg g^{-1}) for a dosage of 0.25 g/100 ml of 100 mg L^{-1} dye solution over a period of 180 min. It was observed that the surface of the adsorbent was negatively charged at higher pH of 11, thus favoring adsorption through electrostatic force of attraction.

Zhang et al. [35] observed that use of brown-rot fungi from decayed pine tree root has increased adsorption capacity by 19.45% (42.63 mg g^{-1}) compared to the pine saw dust. They have observed meso-thermal range of temperature (40 °C) and near-neutral pH of 8 for maximum removal efficiency (99.4%) over a period of 24 h. Due to the presence of plenty of lignocelluloses, it is found that bio-sorbent made from degradation of brown-rotted fungi *P. cocos* was very efficient for the removal of MG from wastewater.

Another sustainable way of adsorption was demonstrated by Srivastava and Rupainwar [58] by using powdered form of tree barks (Neem and Mango). Just with initial washing, drying, and grinding, these materials were found to exhibit relatively good adsorptive tendency (8.75 mg g^{-1} and 12.88 mg g^{-1}, respectively) under the specified experimental conditions (temperature 40 °C and dosage of 0.5 g/50 ml). However, there is some difference in the optimum pH and contact time between these two materials, owing to the difference in their specific surface area and presence of function groups. Sharma et al. [41] observed enhanced adsorption capacity (500 mg g^{-1}) for chemically modified Neem bark using 37% formaldehyde and 0.2 N sulfuric acid under an extreme acidic pH of 2 at 27 °C with a dosage of 0.25 g/100 ml for varying initial concentrations.

In an interesting experiment, banana pseudo-stem fiber was used as adsorbent (adsorption capacity of 26.51 mg g^{-1}) for removing MG (10–40 mg L^{-1}) under nominal environmental conditions (pH 7.0, temperature 25 °C) with a dosage of 0.1 g/50 ml of dye solution [42]. Gupta et al. [43] studied the adsorptive potential of Ashoka (*Saraca asoca*) tree leaves after washing, drying, and grinding. It is found throughout India, especially in Himalaya, Kerala

Bengal, and whole south region. They have observed high adsorptive capacity (83.3 mg g^{-1}) under the following conditions: pH 6, temperature 30 °C, dosage 0.1 g/50 ml, but with low initial concentration of MG (10 mg L^{-1}).

Makeswari and Santhi [39] developed adsorbent from epicarp of *Ricinus communis* subjected to microwave assisted chemical activation using zinc chloride. Orthogonal array experimental design method was used to optimize the preparation of adsorbent. The optimum conditions were found to be microwave power of 100 W, microwave radiation time of 4 min, concentration of zinc chloride of 30% by volume and impregnation time of 16 h. Although the preparation involved many steps, the maximum adsorption capacity was only 12.65 for microwave-treated and 24.39 for zinc chloride activated adsorbents.

Stem and leaf powders of *Daucus carota* plant was investigated by Kushwaha et al. [40] under the following conditions: pH 7.0, temperature 30 °C, dosage 0.5–2.0 g L^{-1}, and time 30 min. The maximum adsorption capacity was observed to be 43.4 mg g^{-1} for stem and 52.6 mg g^{-1} for leaf, which is remarkably good since there is no pre-treatment required. Dahri et al. [57] investigated the potential of *Casuarina equisetifolia* needle, which is widely available, cheap, and ready to use without much pre-treatment. The results showed that adsorption capacity was high (77.6 mg g^{-1}) at pH of 5.2, temperature 25 °C with dosage of 0.08 g/20 ml for 180 min in contact in dye solution having initial concentration 100 mg L^{-1}. After the harvesting the tubers from tuber crops, rest of the parts remain as agricultural waste. *Solanum tuberosum* (potato) plant is an herbaceous perennial of the *Solanaceae* family. Gupta et al. [45] successfully implemented powdered form of potato stem and leaves for the removal of MG. The maximum adsorption capacity was 33.3 mg g^{-1} for leaves and 27 mg g^{-1} for stem under nominal environmental conditions (pH 7.0, temperature 30 °C) for an adsorbent dosage of 0.1 g/50 ml for 33 min contact time with dye at initial concentration of 10–50 mg L^{-1}, which shows remarkable proof for its future up-gradation.

Odoemelam et al. [75] studied the removal of malachite green dye from aqueous solution using neem leaves. Batch adsorption studies were conducted and the adsorption process favored Langmuir and Temkin isotherms. A pseudo-second-order kinetics was favored although the surface adsorption and intraparticle diffusion was identified to be rate-limiting. Krishna Murthy et al. [71] used coffee husk treated with sulfuric acid for the batch adsorptive removal of MG from synthetic dye solution. The Sips isotherm model and pseudo-second-order kinetic model fitted best with experimental data. The diffusion model analysis indicated that the adsorption process was governed by both film and intra-particle diffusion and the study reported that the activated coffee husk was a promising and efficient low-cost adsorbent for the removal of MG from industrial wastewater.

8.2.2   Materials from industrial solid wastes

Due to the increasing energy-demand and accelerated industrialization, developing countries are still depending on the thermal power plants enduring on supply of coal and other fuels. Fly ash and bottom ash are the two main waste materials generated from these power plants (varies from 10% to 50% based on the quality of raw material), which pose serious environmental and economic threats for their safe and sustainable reuse/disposal. Use of industrial waste materials as effective adsorbents proved to be a promising sustainable alternative for conventional waste disposal and treatment. Although many emerging waste materials are being expelled out, researchers in the past two decades showed specific attention to bottom ash, fly ash, and slag materials from various industries (Table 8.2).

8.2.2.1   Fly ash

The current annual worldwide production of coal ash is estimated about 700 million tonnes of which at least 70% is fly ash. It contains 2–5% unburn carbon and has been successful in adsorbing dyes from aqueous solutions. There are a few studies in the last decade reporting the successful modification of coal based fly ash for the removal of reactive dyes from textile effluents [65,66]. Chowdhury and Saha [67] investigated adsorption of MG onto alkali-treated fly ash by blending raw fly ash with 10% lime for 3 h. The enhanced adsorption capacity (4.94 mg g^{-1}) was observed at pH 7, temperature 30 °C, initial concentration 20 mg L^{-1}, contact time 40 min and adsorbent dosage of 1.0 g/100 ml. It was also observed that as the initial dye concentration increased from 10 mg L^{-1} to 100 mg L^{-1}, the adsorption capacity increased initially, then reached equilibrium within 40 min and after that, independent of the initial dye concentration. According to Khan et al. [69], maximum adsorption capacity (0.644 mg g^{-1}) for fly ash was observed at an initial concentration of 5.0 mg L^{-1} for 80 min contact time at pH 7.0, temperature 30 °C and dosage of 1.0 g L^{-1}. Chemical analysis showed that fly ash contained silica as the major constituent, followed by alumina and hematite.

Bagasse fly ash is a similar waste product collected from the particulate separation equipment attached to the flue gas line of the sugarcane bagasse-fired boilers and has been used as an effective adsorbent for the removal of organic dyes. Mall et al. [72] reported maximum adsorption capacity of 170.33 mg g^{-1} under the following experimental conditions: pH 7–11, temperature 30 °C, dosage 1.0 g L^{-1}, contact time 240 min and initial dye concentration 10–250 mg L^{-1}.

8.2.2.2   Bottom ash

Bottom ash was successful in removing reactive dyes such as tetrazine, erythrosine, methyl orange, and metanil yellow [82]. They have attained nearly 96% saturation during the exhaustion in column experiments and 88% of the

Table 8.2 Experimental details and results of MG adsorption using agricultural materials

Adsorbent	Experimental conditions					Isotherm model	Isotherm parameters	Kinetic model	Kinetic parameters ($g\ mg^{-1}$ min)	Thermodynamic parameters ($kJ\ mol^{-1}$)	Overall removal efficiency (%)	Max. adsorption capacity ($mg\ g^{-1}$)	Remarks	References
	pH	Dosage ($g\ L^{-1}$)	Time (min)	Temperature (°C)	IC of MG									
Parts from fruits and grains														
De-oiled soya	5	2	1440	30	1×10^{-4} M to 1×10^{-5} M	F	KF=0.25; n=0.03	PFO	k=0.013	$-\Delta G=5.15$ $\Delta H=27.96$ $\Delta S=103.7$	100	20.44	97% regeneration with acetone in five cycles; film diffusion-limited	Mittal et al. [32]
Wheat bran	5	100	40	28	12–348 $mg\ g^{-1}$	F	kF=0.27; n=1.75				90	24	Adsorption and degradation by the extracellular ligninolytic enzymes	Papinutti et al. [56]
Lemon peel		1.67	1440	32	100 mg L^{-1}	L, R_P	Qo=51.73; KL=0.06 A=3.02; B=0.06					51.73	Non-linear form of isotherm model	Kumar [80]
Orange peel		0.3	1440	32	50–200 mg L^{-1}	L, R_P	Qo=483.63; KL=0.10 A=52.27; B=0.11					483.63	Single stage batch adsorber was designed for different operating line (V/M) ratios	Kumar and Porkodi [46]
Rice bran	6	10	60–90	20	80 mg L^{-1}	F	kF=8.25; n=1.68	PSO	k=0.267 g mg^{-1} min^{-1}	$-\Delta G=35.0$		68.97	External diffusion and intra-particle diffusion sequentially controlled rate	Wang et al. [55]
Wheat bran	6	10	60–90	20	80 mg L^{-1}	F	kF=2.37; n=1.53	PSO	k=0.037 g mg^{-1} min^{-1}	$-\Delta G=30.8$		66.57	External diffusion and intra-particle diffusion sequentially controlled rate	Wang et al. [55]

Adsorbent	Experimental conditions					Isotherm model	Isotherm parameters	Kinetic model	Kinetic parameters ($g\ mg^{-1}\ min$)	Thermo-dynamic parameters ($kJ\ mol^{-1}$)	Overall removal efficiency (%)	Max. adsorption capacity ($mg\ g^{-1}$)	Remarks	References
	pH	Dosage ($g\ L^{-1}$)	Time (min)	Temperature (°C)	IC of MG									
Maize cob (Zea maize)	8	12	25	27	20–150 $mg\ L^{-1}$	L, F	Q_o=11.89; b=0.43; K_L=0.10; k_F=11.04; n=0.59	PSO	k=0.075 g mg^{-1} min		96.2	80.64	Two staged rate external mass transfer followed by intraparticle diffusion	Sonawane and Shrivastava [52]
Mango seed husks	5	5	120	30	100–500 mg L	F	k_F=6.547; n=2.78	PSO	k=0.005	-ΔG=3.93 -ΔH=50.4 -ΔS=0.15 J mol^{-1} K^{-1}	88–97	47.9	Spontaneous, exothermic	Franca et al. [53]
Ginger waste	9	0.2	150	30–50	5–20 $mg\ L^{-1}$	L, F	Q_0=84.03 $mg\ g^{-1}$; b=0.01(l/mg); K_L=0.826 k_F=1.14; n=.93	PSO	k=1.11 g mg^{-1} min^{-1}	-ΔG=1.515 ΔH=47.49 ΔS=0.167		188.6	Desorption and regeneration 40–60%; Chemisorption	Ahmad and Kumar [23]
Degreased coffee bean	10–12	2	240	25–45	25–100 $mg\ L^{-1}$	L, F	Q_o=55.3 mg g^{-1}; K_L=0.09 k_F=2.03; n=0.51	PFO, PSO	k1=0.19; k2=0.32	-ΔG=10.0 ΔH=27.2 ΔS=32.6	98	23	Desorption using 0.1 M HCl in three cycles; chemisorption; spontaneous and endothermic	Baek et al. [24]
Tamarind fruit shell	5	10	60	30	10–200 $mg\ L^{-1}$	L	Q_o=1.95; K_L=146.5	PSO	k=1.77	-ΔG=44.9 ΔH=348.07 ΔS=1.296	98.8	1.951	Thermodynamically favorable	Saha et al. [33]
Almond shell	4.5	10	60	40	70–1150 $mg\ L^{-1}$	L, F	Q_0=29.9; b=0.0057; R_L=0.132 k_F=1.365; n=2.3	PSO	k=0.09 g mg^{-1} min^{-1}	-ΔG=2.87 ΔH=21.67 ΔS=83.47 J mol^{-1} K^{-1}		29	Regeneration with ethanol 69% in five cycles; Spontaneous, endothermic	Ozdes et al. [49]

Adsorbent	Experimental conditions					Isotherm model	Isotherm parameters	Kinetic model	Kinetic parameters (g mg⁻¹ min)	Thermodynamic parameters (kJ mol⁻¹)	Overall removal efficiency (%)	Max. adsorption capacity (mg g⁻¹)	Remarks	References
	pH	Dosage (g L⁻¹)	Time (min)	Temperature (°C)	IC of MG									
Modified rice husk	7		120	40	10–100 mg L⁻¹	F	$kF=7.395$; $n=1.92$	PSO	$k=0.119$ g mg⁻¹ min⁻¹	$-\Delta G=6.18$ $\Delta H=63.76$ $\Delta S=234.4$ J mol⁻¹ K⁻¹	98.9	17.98	Spontaneous, feasible, endothermic	Chowdhury et al. [59]
Cucumis sativa fruit peel	6	20	80	27		L	$Qo=36.23$; $KL=0.07$	PSO	$k=0.015$ g mg⁻¹ min⁻¹		99.86	36.23	Tested for real effluent	Santhi and Manonmani [54]
Bivalve shell (BS) and *Zea mays* L. husk leaf (ZHL)	6	0.5	30	50	10–200 mg L⁻¹	L	$Qo=81.5$; $KL=0.45$	PFO, PSO	$k1=0.83$ $k2=0.80$	$-\Delta G=5.93$ $\Delta H=32.4$ $\Delta S=127$	98	81.5	Boundary layer diffusion limited (Physisorption)	Jalil et al. [29]
Potato peel	12	2.5		27	5–25 mg L⁻¹	F	$kF=3.5$; $n=0.98$	PSO	$k=0.01$ g mg⁻¹ min⁻¹		82.7	111		Sharma et al. [41]
Walnut shell	5	1.5	120	25	20–600 mg L⁻¹	L, F, R-P, Sips	$Qo=90.8$; $b=0.14$; $KL=0.09$ $kF=21.6$; $n=3.67$ $KR=47.9$; $B=0.84$; $aR=1.29$ $Qm=116$; $ks=0.2$; $KLF=0.52$	PFO, PSO	$k1=0.01$; $k2=0.002$	$-\Delta G=7.5$ $\Delta H=8.2$ $-\Delta S=52.4$	87	90.8	Desorption using 0.1 M NaOH for five cycles; Spontaneous and endothermic	Dahri et al. [25]
Wood apple shell	7–9	0.4	210	26	100–700 mg L⁻¹	L	$Qo=35.84$; $KL=0.033$	FO	$K=0.053$	$-\Delta G=2.83$ $\Delta H=1.58$ $\Delta S=6.37$	98.8	80.645	Spontaneous and endothermic	Sartape et al. [34]
Pea shells	7	10	35	30	10–70 mg L⁻¹	F	$kF=0.58$; $n=1.64$	PSO	$k=0.133$ g mg⁻¹ min⁻¹	$-\Delta G=7.03$ $\Delta H=125.73$ $\Delta S=445$	96	14.5	Spontaneous and endothermic	Khan et al. [79]

Adsorbent	Experimental conditions					Isotherm model	Isotherm parameters	Kinetic model	Kinetic parameters (g mg^{-1} min)	Thermo-dynamic parameters (kJ mol^{-1})	Overall removal efficiency (%)	Max. adsorption capacity (mg g^{-1})	Remarks	References
	pH	Dosage (g L^{-1})	Time (min)	Temperature (°C)	IC of MG									
Breadnut peel	3.6	2	120	25	0–1000 mg L^{-1}	Sips	Qo=180.0; Ks=0.008; n=1.19	PSO	k=0.32 g mg^{-1} min^{-1}	-ΔG=7.04 -ΔH=11.74 -ΔS=16.09 J mol^{-1} K^{-1}	76	353	Regeneration with 250 ml of 0.1 M HNO$_3$ >90% in five cycles; Spontaneous, endothermic, exothermic for modified sorbent	Chieng et al. [48]
Types of saw dust														
Rosewood saw dust	6–9	4	120	26	50–250 mg L^{-1}			FO	K=0.01		99.5	26.9 and 74.5	Low pH favors	Garg et al. [26]
Prosopis Cineraria sawdust	6–10	4	180	27	0–500 mg L^{-1}			FO	0.02–0.03 min^{-1}		99.1	65.8	Lower adsorption efficiency than GAC at higher dye concentrations	Garg et al. [62]
Rubber wood sawdust				32	100 mg L^{-1}	L, R_P	Qo=33.25; KL=0.21 A=1498; B=0.97					36.45	Non-linear form of isotherms	Kumar and Sivanesan [51]
Rattan saw dust	4	1.5	210	30	25–300 mg L^{-1}	L, F	Qo=62.7; b=0.02 KF=5.88; n=2.41	PFO PSO	k1=0.14 k2=22.2			62.71	Combination of film and pore diffusion control rate	Hameed and Khaiary [27]
Neem saw dust	7.2	5	14	25	6–12 mg L^{-1}	L	Qo=4.35; b=1.31; KL=0.08	FO	k=0.09	-ΔG=4.02 -ΔH=54.56 -ΔS=169.57	85	4.35	Regeneration with 1.5% KCl; dipolar interactions	Khattari and Singh [30]
Beech sawdust	5	2.5	120	20	10–1000 mg L^{-1}	RP	k=1.36; a=0.008; b=1.106	PSO	k=0.96 g mg^{-1} min^{-1}	-ΔG=6.78 ΔH=36.04 ΔS=140.5 J mol^{-1} K^{-1}		83.21	Spontaneous, feasible and endothermic; multi-stage sorption	Witek-Krowiak [44]

Adsorbent	Experimental conditions					Isotherm model	Isotherm parameters	Kinetic model	Kinetic parameters (g mg⁻¹ min)	Thermo-dynamic parameters (kJ mol⁻¹)	Overall removal efficiency (%)	Max. adsorption capacity (mg g⁻¹)	Remarks	References
	pH	Dosage (g L⁻¹)	Time (min)	Temperature (°C)	IC of MG									
Camphora saw dust	9.3	1	720	60		L	$Q_o=282.5$; $RL=0.008$	PSO	$k=0.365$ g mg^{-1} min^{-1}	$-\Delta G=540.77$ $\Delta H=4868.03$ $\Delta S=14.52$	96.7	280.3 for oxalic acid-treated, 222.8 for citric acid-treated, and 157.5 tartaric acid-treated	Spontaneous and feasible; biochar used for real effluent	Wang et al. [38]
Other plant parts														
Rice straw	4	1.5	120	20	100–500 mg L⁻¹	L, F	$Q_o=94.34$; $b=0.064$ $kF=29.07$; $n=4.78$	PFO			93	256.41	Intraparticle diffusion rate increased	Gong et al. [47]
Oil palm trunk fibre	8	1.5	120	30	25–300 mg L⁻¹	Multi-layer isotherm	$Q_o=50.6$; $k1=0.07$; $K2=0.003$	PFO	$k=0.03$			149.35	External mass transfer limited	Hameed and Khaiary [28]
Dead leaves of plane tree	5.5	2.5	1440	45	5–50 mg L⁻¹	L	$Q_o=85.47$; $b=31.23$	PSO	$k=2.91$	$-\Delta G=25.46$ $\Delta H=5.7$ $\Delta S=104.5$		89.43	Spontaneous and endothermic	Hamdaoui et al. [36]
Firmiana simplex wood fiber (Phoenix tree)	11	2.5	180	25	100 mg L⁻¹	F	$kF=0.288$; $n=1.08$	PSO	$k=6.4$	$-\Delta G=3.65$ $\Delta H=61.4$ $\Delta S=234.3$		142.4	Spontaneous and endothermic	Pan and Zhang [37]
Pine tree root decayed by brown-rot fungi (BRW)	8	2	1440	40	100 mg L⁻¹	L	$Q_o=25.78$; $KL=0.43$	PSO	$k=0.74$	$-\Delta G=27.11$ $\Delta H=40.82$ $\Delta S=232.59$	99.4	42.63	75% regeneration with acetone in three cycles; chemisorption; spontaneous and endothermic	Zhang et al. [35]

Adsorbent	Experimental conditions					Isotherm model	Isotherm parameters	Kinetic model	Kinetic parameters (g mg⁻¹ min)	Thermodynamic parameters (kJ mol⁻¹)	Overall removal efficiency (%)	Max. adsorption capacity (mg g⁻¹)	Remarks	References
	pH	Dosage (g L⁻¹)	Time (min)	Temperature (°C)	IC of MG									
Neem bark	5	10	120	40	1×10^{-4} M to 1×10^{-6} M	L	Q_o=4.88e4 mol g⁻¹; b=21.46e-3 L mol⁻¹	PSO	k=13.19 g mol⁻¹ min⁻¹	$-\Delta G$=21.45 ΔH=20.41 ΔS=180 J mol⁻¹ K⁻¹	88.45	8.75	Spontaneous, endothermic	Srivastava and Rupainwar [58]
Mango bark powder	2	10	150	40	1×10^{-4} M to 1×10^{-6} M	L	Q_o=5.17e4 mol g⁻¹; b=6.28e-3 L mol⁻¹	PSO	k=29.06 g mol⁻¹ min⁻¹	$-\Delta G$=24.36 ΔH=25.35 ΔS=130 J mol⁻¹ K⁻¹	99.45	12.88	Spontaneous, endothermic	Srivastava and Rupainwar [58]
Banana pseudo-stem fibers	7	2	60	25	10-40 mg L⁻¹	L, F	Q_o=26.52; b=0.09 kF=2.66; n=1.47	PSO	k=0.144	$-\Delta G$=3.09; $-\Delta H$=11.7 $-\Delta S$=27.14 J mol⁻¹ K⁻¹	82.6	26.51	Spontaneous, exothermic	Gupta et al. [42]
Ashoka leaf powder	6	2	25	30	10 mg L⁻¹	F	kF=2.83; n=1.14	PSO	k=0.16 g mg⁻¹ min	$-\Delta G$=4.19 $-\Delta H$=6.67 $-\Delta S$=8.19 J mol⁻¹ K⁻¹	85	83.3	Regeneration with 250 ml of 0.1 M HCl–ethanol >90% in five cycles; Spontaneous, exothermic	Gupta et al. [43]
Epicarp of *Ricinus communis*	5	4	120	30	100 mg L⁻¹	L	Q_o=12.65; b=0.204	PSO	k=0.067 g mg⁻¹ min⁻¹			12.65 for MRC (Single) 11.76 for MRC (binary) 24.39 for ZRC (single) and 20.41 for ZRC (binary)	Competitive adsorption in binary system	Makeswari and Santhi [39]

Adsorbent	Experimental conditions					Isotherm model	Isotherm parameters	Kinetic model	Kinetic parameters (g mg⁻¹ min)	Thermodynamic parameters (kJ mol⁻¹)	Overall removal efficiency (%)	Max. adsorption capacity (mg g⁻¹)	Remarks	References
	pH	Dosage (g L⁻¹)	Time (min)	Temperature (°C)	IC of MG									
Stem and leaf powders of *Daucus carota*	7	2	30	30		L, F	Q_o=55.5; b=0.044; kF=2.60; n=1.23	PSO	k=0.34	$-\Delta G$=3.82 $-\Delta H$=7.49 $-\Delta S$=12.13 J mol⁻¹ K⁻¹	82	43.4 for stem and 52.6 for leaf	Spontaneous, exothermic; surface and pore diffusion	Kushwaha et al. [40]
Rice straw	8	10	20	35	50 mg L⁻¹	L, F		FO	k=9.4 g mg⁻¹ min⁻¹	$-\Delta G$=7.7 ΔH=436.83 ΔS=430 J mol⁻¹ K⁻¹	86.1	92	Regeneration with 0.1 M NaOH 99% in three cycles; Spontaneous, endothermic	Singh and Kaur [50]
Neem bark	2	2.5		27	5–25 mg L⁻¹	F	kF=9.4; n=1.14	PSO	k=0.36 g mg⁻¹ min⁻¹		94.4	500		Sharma et al. [41]
Casuarina equisetifolia needle	5.2	4	60	25	100 mg L⁻¹	L	Q_o=77.6; b=0.072; KL=0.027	PSO	k=1.8 g mg⁻¹ min⁻¹	$-\Delta G$=4.43 ΔH=33.8 ΔS=127.9 J mol⁻¹ K⁻¹	88	77.6	Spontaneous, feasible, endothermic	Dahri et al. [57]
Potato plant waste	7	2	33	30	10–50 mg L⁻¹	L, F	Q_o=41.6; b=0.048; kF=2.27; n=1.3	PSO	k=0.75	$-\Delta G$=3.36 $-\Delta H$=9.08 $-\Delta S$=18.9 J mol⁻¹ K⁻¹	67–87	33.3 for leaves and 27 for stem	Regeneration with 1% HCl/ethanol solution 96% with three cycles; Spontaneous, exothermic	Gupta et al. [45]

dye material was recovered by eluting with diluted sodium hydroxide solution. Wang and Li [76] separated unburned carbon from fly ash and been employed as a low cost adsorbent for a basic dye adsorption (Rhodamine B) in aqueous solution. Mane et al. [73] reported the removal of brilliant green using bagasse bottom ash.

In a similar study, Gupta et al. [20] attempted to explore the possibility of adsorption of MG using bottom ash from thermal power plant. The dried material was treated with hydrogen peroxide to oxidize the adhering organic material, followed by successive heating in oven (100 °C) and muffle furnace (500 °C for 15 min in presence of air). The maximum adsorption capacity (0.71 mg g^{-1}) was observed under the following experimental conditions: pH 5.0, temperature 30 °C, dosage 0.1 g/25 ml, contact time 60–120 min, and initial concentration 1×10^{-5} M to 1×10^{-4} M. It was reported that the low adsorption of dye in lower pH range is due to the development of positive charge at bottom ash particles, while in basic medium the formation of electric double layer changes its polarity and consequently the dye uptake increased. Bottom ash contains metal ion oxides and hydroxides, which form hydrogen bonds with the MG [20]. Batch and column experiments revealed that quantitative recovery of the dye can be achieved by eluting acetone through the column and adsorbent can be regenerated (Table 8.2).

8.2.2.3   Other industrial solid waste materials

Gupta et al. [77] reported the use of blast furnace slag for removal of MG by activating it with hydrogen peroxide followed by washing with 0.1 N hydrochloric acid and drying at 100 °C. Activated slag was also prepared from this by heating in a muffle furnace at 600 °C for 1 h with air. The uptake of MG on activated carbon and slag increased with temperature indicating the endothermic nature of the adsorption process. The maximum adsorption capacity was 8×10^5 mol g^{-1} under the optimum conditions of pH 6 and temperature of 40 °C. Even while considering the extensive procedure for activation, high amount of dosage requirement (10 g L^{-1}), significantly long contact time (360–480 min), and the range of initial dye concentrations tested (1×10^{-4} M to 1×10^{-3} M), the performance of the adsorbent was substantiated mainly based on its high adsorption capacity. Khan et al. [70] studied adsorptive removal of MG using untreated Akash Kinari coal (from Coal India) having carbon, silica, and alumina as their major constituents. The observed adsorption capacity was very small (0.8 mg g^{-1}) considering the prevailing experimental conditions: pH 7.2, temperature 20–40 °C, dosage 1 g/50 ml, contact time 110 min and initial concentration of dye 5–20 mg L^{-1}.

Two paper mill waste materials (de-inking paper sludge and organic sludge from virgin pulp mill) were used as adsorbent precursors for removal of MG from aqueous solution [63]. Adsorbents were prepared by taking 20 g of paper mill waste material in a covered ceramic cup placed in a nickel recipient and subjected to pyrolysis in an electrical furnace at 650 °C for 2 h. Maximum adsorption capacity was observed for the earlier one (982 mg g^{-1}) than for latter (435 mg g^{-1}). Apart from this, thermal decomposition of MG was influenced by the nature of dispersion on the adsorbent surface.

Another critical industrial solid waste is red mud, which is formed after the caustic digestion of bauxite ores during the production of alumina. The waste red mud was employed for the removal of Congo Red from wastewater. Acid treatments have been used to neutralize the alkalinity of red mud (pH of 10.0–12.5), which could also be recognized as a method for activating the material. Acid-activated sintering process red mud was investigated as an adsorbent for removal of MG [78]. It was blended with 0.5 M hydrochloric acid at a solid-to-liquid ratio of 1/20 for 0.5 h at 25 °C. The dye adsorption capacities increased from 2.4 mg g^{-1} to 93.2 mg g^{-1} for MG with the concentrations ranging from 10 mg L^{-1} to 500 mg L^{-1}. Since the pH zero point charge (pH$_{zpc}$) of the adsorbent was 3.2, the cationic dye molecules repelled at the surface of the adsorbent when the pH value was less than 3.2 (resulted in the low adsorption) and when the pH was greater than 3.2, negative charges were dominant on the surface (facilitated the adsorption). Based on the high affinity and capacity, fly ash can be suggested as a promising sustainable adsorbent. However, the decreasing trend of research on industrial waste products indicates the ample evidence of emerging smart adsorbents. Gurkan and Semra [64] examined the adsorption of malachite green dye from aqueous solution on waste foundry sand. The effect of contact time, adsorbent amount, pH, and initial dye concentration were investigated. The adsorption data was well fitted by the pseudo second order model.

8.2.3   Influence of initial conditions

8.2.3.1   Effect of pH

Many researchers have observed that the adsorption of positively charged dye molecules was enhanced at higher solution pH [26,27,30]. This can be generally explained on the basis of change in electrostatic forces of attraction. At acidic pH, hydrogen ions may compete with dye ions for the adsorption sites of adsorbent, thereby inhibiting the adsorption of dye. At higher solution pH, the surface of the adsorbent may get negatively charged, which enhances the adsorption of positively charged dye. Also a change of solution pH affects the adsorptive process through dissociation of functional groups on the adsorbent surface. This was particularly significant for the adsorption onto neem saw dust where the surface of cellulose in contact with water was negatively charged [30].

In contrary to this, Chieng et al. [48] observed that percentage removal of MG increased by 76% when the pH increased from 2 to 3.6 (ambient) which implies that adsorption capacity could be directly influenced by change in interactive force between either H^+ or OH^- ions (Table 8.1). However, at higher pH, the amount of MG being removed remained almost constant. Similar trend has been reported for alkali-modified rice husk [59] and maize cob [52].

Khan et al. [69] suggested that the variation of adsorption with pH is also dependent on the difference in the molecular structure of its main constituents (silica and alumina) as indicated by their pH_{zpc} values (2.3 and 8.2, respectively). When the solution pH is below the pH_{zpc} of fly ash (5.8), the surface would have high positive charge density and it will be reversed at high pH values. Khan et al. [70] observed that increase in pH from 3.2 to 7.2 resulted in an increase in adsorption capacity from 62.16 to 83.42 by the coal. Among the adsorption studies onto fly ash, Mall et al. [72] reported highest adsorption capacity (170.33 mg g^{-1}) for bagasse fly ash although initial conditions like pH of 7 and temperature of 30° were same in all coal based adsorbents [68,69]. Among the industrial adsorbents, only paper mill sludge materials [63] required alkaline pH (9.2–10.3) whereas activated sintering red mud [78] required most acidic pH (3.2) for optimum performance (Table 8.2).

8.2.3.2   Effect of initial dye concentration

In general, it was observed that increasing the initial dye concentration results in an increase in the adsorption capacity because it provides a driving force to overcome all mass transfer resistances of dyes between the aqueous and solid phases. However, the sorption percentage decreased with increase in initial dye concentration. This is due to the continuous filling of the active sites on the adsorbent surface by dye molecules, which become saturated at a certain concentration. This postulates the importance of mass transfer derived from increased concentration gradient with the increase of initial dye concentration. However, this can inversely impact the adsorption frequency because of the limited adsorption sites available for further uptake of cationic dye. In comparison with commercially available activated carbon, the carbon prepared from saw dust had lower adsorption efficiency at higher dye concentrations [26]. The optimum pH for dye removal was found be over the range of 6–9 whereas adsorption efficiency of activated carbon was unaffected by solution pH. It was observed that as the initial dye concentration increased from 10 mg L^{-1} to 100 mg L^{-1}, the adsorption capacity of the treated rice husk increased [59].

Based on the results from batch kinetic experiments, Dizge et al. [66] reported that the amount of reactive dyes adsorbed onto fly ash increased with increasing initial dye concentration and contact time, although for some of the reactive dyes, pH was a major influencing factor. Khan et al. [69] reported that the removal of MG increased from 0.219 mg g^{-1} to 0.644 mg g^{-1} by increasing the initial dye concentration from 5 mg L^{-1} to 20 mg L^{-1} as well as by increasing the contact time (till reaching optimum of 80 min). Based on the selected literature, it was observed that highest adsorption capacity was reported when initial concentration of MG is higher, irrespective of the dosage of adsorbents (Table 8.3). Among the adsorbents prepared from industrial waste origin, the highest adsorption capacity was reported for paper mill waste sludge, bagasse fly ash and activated sintered red mud (982 mg g^{-1}, 170.33 mg g^{-1}, and 93.2 mg g^{-1}, respectively) where the range of initial concentrations tested were very high (10–4000 mg L^{-1}, 10–250 mg L^{-1}, and 10–500 mg L^{-1}, respectively), while the adsorbent dosage was moderately low (2.5 g L^{-1}, 1.0 g L^{-1}, and 5 g L^{-1}, respectively). The minimum contact time (40 min, 60 min, and 80 min) with highest removal efficiency (99.07%, 96%, and 87%, respectively) was observed for fly ash adsorbents even though their adsorption capacities (4.94 mg g^{-1}, 0.71 mg g^{-1}, and 0.644 mg g^{-1}, respectively) were comparatively low [20,59,69].

8.2.3.3   *Effect of ionic strength of solution*

Wang et al. [38] stated that increasing ionic strength of the dye solution (by adding KCl from 0.001 M to 0.2 M) decreased the adsorption of MG. This might be due to the competition of the potassium ions and the cationic dye molecules for sites available in sorption process. At higher pH, the surface of adsorbents become negatively charged according to their respective pH$_{zpc}$ values. The chloride ions may have no effect on the adsorption process because of the electrostatic repulsive force at the surface. However, a partial neutralization of the negative surface charge may be occurred due to the presence of potassium ions, resulting in compression of electrical double layer and thereby reducing the attractive forces.

8.2.3.4   *Effect of temperature*

It is shown that adsorption of MG increased with increasing temperature of the solution, suggesting that the adsorption is an endothermic process [38]. This can be attributed to the increase in mobility of the dye molecules, increase in the number of active sites for the adsorption, increased number of dye molecules acquiring energy to interact with active sites as well as a swelling effect within the internal structure of adsorbent enabling large dye molecule to penetrate. For the industrial adsorbents, most of the experiments were conducted at normal temperature (25–30 °C) while the effect of temperature was studied only in a few experiments (40–50 °C) [68,70,77].

Table 8.3   Adsorbent characterization studies

Adsorbent	Pre-treatment	Activation method, if any	Particle size (mm)	Surface area (m² g⁻¹)	Pore volume (cc g⁻¹)	Pore size (Å)	Bulk density (kg m⁻³)	Functional groups identified, if any	Instruments used	Reference
Industrial waste materials										
Blast furnace slag	Hydrogen peroxide washing with 0.1 N HCl, drying	450 °C for 60 min	0.15–0.2	107				Lime, silica, alumina	FTIR, XRD, SAA, SEM	Gupta et al. [77]
Power plant-bottom ash	Hydrogen peroxide	500 °C for 15 min	0.08–0.1							Gupta et al. [20]
Municipal solid waste incinerated-bottom ash	Washing, screening	Thermal activation at 500 °C for 6 h	0.15–0.2	14.10				Alumina-silicates and iron		Gupta et al. [20]
Power plant-generated bottom ash	Washing, screening	Thermal activation at 500 °C for 15 min		0.87	0.46		630.1	Alumina, gypsum, beaverite, borax, kaolinite		Gupta et al. [20]
Akash kinari coal	Nil	Nil	4.9	12.34	0.43		3310	Carbon, silica, alumina	SEM, XRD	Khan et al. [69]
Bagasse fly ash	Washing with hot water (70 °C)		0.18–1.18	168.83	0.053	23.97	133.33	Free and hydrogen bonded OH groups	XRD, SEM, SAA, FTIR	Mall et al. [72]
Fly ash	Nil	110 °C for 24 h	0.075	40.16	0.43		3.51	Surface hydroxylation of oxides	SAA, SEM	Khan et al. [69]
De-inking paper sludge and virgin pulp mill	Chemical	540 °C for 4 h		88.4 and 274	0.014 and 0.061	45.3 and 21.6			TG, DTG, and DTA analysis	Mendez et al. [63]
Fly ash	Lime treatment	70 °C for 24 h; pH 7		112.5	0.087	74			XRD, SEM, SAA	Chowdhury and Saha [67]
Activated sintering-red mud	Acid activation			130	0.231	24.0		Hydrogen bonding between (N) and (OH or siloxane group)		Zhang et al. [78]
Agricultural waste materials										
Ginger waste	Washing, drying at 60 °C for 12 h	Soaked in H₂SO₄ and ZnCl₂ (1:1) for 24 h	0.08–0.15						SEM, FTIR, spec	Ahmad and Kumar [23]
Degreased coffee bean	Washing, screening, drying at 105 °C for 24 h	0.01–5 M NaOH for 12 h		142.1				S, carboxyl linkage	SEM, FTIR	Baek et al. [24]

Adsorbent	Pre-treatment	Activation method, if any	Particle size (mm)	Surface area ($m^2 g^{-1}$)	Pore volume ($cc\ g^{-1}$)	Pore size (Å)	Bulk density ($kg\ m^{-3}$)	Functional groups identified, if any	Instruments used	Reference
Walnut shell	Soaking, washing, drying at 75 °C for 48 h, grinding		<3.55						UV–vis spec, Elemental CHNSO Analyzer, FTIR	Dahri et al. [25]
Saw dust	Drying, grinding	Formaldehyde (1%) in 1:5 at 50 °C for 4 h Sulfuric acid in 1:1 at 150 °C for 24 h					1450		Spec	Garg et al. [26]
Rattan saw dust	Washing, drying, crushing	Nil	0.5–1.0						UV–vis spec.	Hameed and Khalary [27]
Oil palm trunk fiber	Drying, grinding, washing	Nil	0.5–1.0	64.44					SEM, FTIR, UV–vis spec	Hameed and Khalary [28]
Bivalve shell (BS) and Zea mays L. husk leaf (ZHL)	Washing, drying, crushing	$Ca(OH)_2$						CaO	SEM, FTIR, XRF, UV–vis spec	Jalil et al. [29]
Prosopis Cineraria sawdust	Formaldehyde and sulfuric acid	80 °C for 24 h	0.3–0.85	376			690			Garg et al. [62]
Neem saw dust	Boiling, washing, drying	Boiling with dilute HCl for 30 min	0.15–0.3					Cellulose, lignin, silica	FTIR, XRD	Khattari and Singh [30]
Lemon peel	Cutting, drying, grinding	Nil								Kumar [80]
De-oiled soya	Washing, drying	H_2O_2	0.3–0.425					Gorthite, corundum, coesite, laumonite	UV–vis spec, porosimeter, SAA, SEM, XRD, FTIR	Mittal et al. [32]
Tamarind fruit shell	Washing, drying	Nil	>0.177					–OH, –CH_2, C=O	SEM, XRD, FTIR	Saha et al. [33]
Wood apple shell	Washing, drying, grinding	Nil	0.355						FTIR, SEM, Elemental CHNSO analyzer, UV–vis spec	Sartape et al. [34]

Adsorbent	Pre-treatment	Activation method, if any	Particle size (mm)	Surface area (m² g⁻¹)	Pore volume (cc g⁻¹)	Pore size (Å)	Bulk density (kg m⁻³)	Functional groups identified, if any	Instruments used	Reference
Pine tree root decayed by brown-rot fungi (BRW)	Drying, washing, drying	Nil	<0.1	5.39		58		Acidic groups (carboxylic and lactonic), nonacidic (ether, quinine and carbonyl) groups, phenol group and anhydride	SEM, EDS, FTIR, SAA, UV–vis spec	Zhang et al. [35]
Dead leaves of plane tree	Washing, drying, crushing, washing, drying	Nil								Hamdaoui et al. [36]
Firmiana simplex wood fiber (Phoenix tree)	Washing, drying		0.1	226.5			1360			Pan and Zhang [37]
Pea shells	Washing, drying at 80 °C for 3 h	0.5 M citric acid at 120 °C for 90 min						Lignocelluloses, carboxyl group	FTIR, SEM	Khan et al. [79]
Camphora saw dust	Washing, drying at 60 °C for 48 h	1.1 M citric acid, oxalic acid and tartaric acid at 30 °C for 3 h	0.075					Carboxylic groups	SEM, FTIR	Wang et al. [34]
Epicarp of *Ricinus communis*	Drying, grinding	Microwave (100 W for 4 min) assisted $ZnCl_2$ (30–60% v/v) activation						Hydroxyl, lactonic, carbonyl, carboxylic and carboxylate groups	SEM, FTIR, EDAX	Makeswari and Santhi [39]
Stem and leaf powders of *Daucus carota*	Washing, drying, grinding	Nil	0.1–0.15					Hydroxyl, methyl and carbonyl groups	SEM, FTIR	Kushwaha et al. [40]
Cucumis sativa fruit peel	Washing, crushing, drying at 110 °C	98% H_2SO_4 for 12 h						Hydroxyl, methyl and carbonyl groups	FTIR	Santhi and Manonmani [54]
Maize cob (*Zea maize*)	Drying, grinding, washing, drying at 50 °C for 12 h	Nil	0.25–0.42	24.19	0.0025	15.77		Hydroxyl, methyl, carbonyl, C=O, C=N and C+C groups	FTIR, XRD, SEM, CHN, SAA	Sonawane and Shrivastava [52]
Rubber wood sawdust	Washing, drying at 60 °C for 48 h	Nil	0.075–0.15							Kumar and Sivanesan [51]
Mango seed husks	Drying, grinding	Nil	<0.43					Phenolic and carboxylic groups		Franca et al. [53]
Potato plant waste	Washing, drying at 60 °C for 45 min, grinding	Nil	0.1–0.15					Hydroxyl, methyl and carbonyl groups	SEM, FTIR	Gupta et al. [45]

Adsorbent	Pre-treatment	Activation method, if any	Particle size (mm)	Surface area (m² g⁻¹)	Pore volume (cc g⁻¹)	Pore size (Å)	Bulk density (kg m⁻³)	Functional groups identified, if any	Instruments used	Reference
Orange peel	Cutting, drying, grinding	Nil	0.33–0.92							Kumar and Porkodi [46]
Rice straw	Cutting, washing, drying at 50 °C for 24 h, grinding	0.5 M citric acid in 1:12 (w/v) for 30 min, followed by suspension in 0.1 M NaOH	0.42–0.84					Hydroxyl and carboxyl groups	FTIR	Gong et al. [47]
Rice straw	Cutting, washing, drying, grinding	Nil	0.35–0.60					Hydroxyl, carbonyl and carboxyl groups	FTIR	Singh and Kaur [50]
Almond shell	Washing, drying at 40 °C for four days, grinding	Nil	<0.15					Hydroxyl, carbonyl and carboxyl groups	FTIR	Ozdes et al. [49]
Breadnut peel	Drying, blending	1 g in 100 ml of 4.0 M NaOH for 2 h	0.355–0.850					Hydroxyl, carbonyl, amine, phenyl and carboxyl groups	FTIR, XRF, SEM	Chieng et al. [48]
Beech sawdust	Nil	Nil	0.053–0.075					Carboxyl, carbonyl, hydroxyl and amino groups		Witek-Krowiak [44]
Ashoka leaf powder	Washing, drying at 60 °C, grinding	Nil	0.10–0.15					Hydroxyl, methyl, carbonyl and carboxyl groups	SEM, FTIR	Gupta et al. [43]
Potato peel	Washing, drying at 65 °C for 24 h, grinding	6% HCHO for 24 h, 0.2 N H₂SO₄ as 1:1 (w/v)	0.08–0.15							Sharma et al. [41]
Neem bark	Washing, drying at 65 °C for 24 h, grinding	37% HCHO and 0.2 N H₂SO₄	0.08–0.15					Cellulosic compounds (carboxylic and phenolic groups)		Sharma et al. [41]
Modified rice husk	Washing, drying at 70 °C for 3 h	5% NaOH autoclaved at 10 psi for 15 min						Hydroxyl, ether, and carboxyl groups	SEM	Chowdhury et al. [59]
Casuarina equisetifolia needle	Washing, drying at 70 °C, blending		355–500					Hydroxyl, carbonyl, amine, and carbonyl groups	FTIR, XRF CHNSO	Dahri et al. [57]
Neem bark	Washing, drying at 110 °C for 24 h, grinding	Nil	0.15–0.212	328.1					SEM	Srivastava and Rupainwar [58]

Adsorbent	Pre-treatment	Activation method, if any	Particle size (mm)	Surface area ($m^2 g^{-1}$)	Pore volume ($cc g^{-1}$)	Pore size (Å)	Bulk density ($kg m^{-3}$)	Functional groups identified, if any	Instruments used	Reference
Mango bark powder	Washing, drying at 110 °C for 24 h, grinding	Nil	0.15–0.212	544.5					SEM	Srivastava and Rupainwar [58]
Rice bran	Dried at 105 °C for 24 h	Nil	0.15–0.25							Wang et al. [55]
Wheat bran	Dried at 105 °C for 24 h	Nil	0.15–0.25							Wang et al. [55]
Wheat bran	Dried at 60 °C for 24 h	Nil	0.8–2.4							Papinutti et al. [56]
Banana pseudo-stem fibers	Cutting, boiling, drying, grinding	Nil	0.2–0.3					Hydroxyl, carbonyl and carbonyl groups	FTIR, XRD, SEM	Gupta et al. [42]

8.3   Activation strategy for low-cost adsorbents

It has been clearly stated that the type of carbon sorbent and its mode of preparation exert a marked influence on the adsorption capacity. Therefore, the adsorption characteristics of lignite-based carbon markedly depended on the mode of activation.

8.3.1   Activated carbon from agricultural solid wastes

Gong et al. [47] thermo-chemically modified rice straw by adding citric acid as an esterifying agent. Two introduced free carboxyl groups of esterified rice straw were further loaded with sodium ion to yield potentially biodegradable cationic sorbent (Table 8.1). In a similar attempt, the adsorption capacity of MG was enhanced when breadnut skin was chemically modified with sodium hydroxide, leading to an adsorption capacity of 353.0 mg g^{-1} that was far superior to most reported adsorbents [48].

8.3.2   Efficient adsorbents from industrial solid wastes

Gupta et al. [77] revealed that presence of anionic surfactants does not affect the uptake of dye significantly. Further, the column capacity for both systems was found to be higher than the batch capacity. Gupta et al. [20] developed a fixed-bed adsorption column with bottom ash as the adsorbent material (Table 8.2). The adsorbent was prepared by physical (washing and drying) and chemical pre-treatment (oxidizing with hydrogen peroxide), followed by thermal activation (at 500 °C for 15 min in presence of air). Mendez et al. [63] reported that maximum adsorption capacity obtained by Langmuir equation was higher for the adsorbent from de-inking paper sludge (982 mg g^{-1}).

8.4   Selection of appropriate mass transfer models

8.4.1   Isotherm models

Gupta et al. [81] observed that both the Freundlich and Langmuir models could be used to represent equilibrium conditions for MG adsorption onto power plant-generated bottom ash (Table 8.2). Dincer et al. [65] found that adsorption of MG onto the coal-based bottom ash was better described by the Freundlich isotherm. It was observed that adsorption with most of the industrial waste generated materials were following Langmuir and Freundlich isotherm models, while bagasse fly ash reported to follow R-P isotherm also [72].

8.4.2   Kinetic models

Most of the kinetic models for bio-sorption from literature rely on goodness of fit towards either molecular diffusion and/or chemical reactions to identify the rate-limiting step and overall rate of reaction. Among the models based on chemical reactions, pseudo-first-order (PFO) and pseudo-second-order (PSO) are most widely employed for studying kinetics of dye adsorption [44]. Most of the researchers have reported that pseudo-second-order kinetic model can represent the best fit relationship which implies that adsorption process is interaction controlled, as well as chemisorption involved [35]. However, the initial phase of adsorption can be physical or ion exchange depending on the surface characteristics of the adsorbents. Therefore, it is to be understood that there are at least two mechanisms involved in the actual process.

Kinetic studies of dye removal with bottom ash confirm the first order process for the adsorption reaction via particle diffusion process (Mittal et al., 2005). Based on the kinetic data, Gupta et al. (2003) showed the uptake of MG on the prepared adsorbents to be of first order. Dincer et al. [65] showed that the dye-stuff uptake process for MG onto the coal-based bottom ash followed the second-order kinetics. Linear plots of t/qt versus t at different temperatures showed that kinetic data can be successfully fitted to the pseudo-second-order (PSO) equation. However, the rate constant decreased as the temperature increased indicating exothermic nature of adsorption of MG to treated fly ash [68].

8.4.3   Thermodynamic models

Based on the mean free energy estimated from the Dubinin-Radushkevich model, Chowdhury and Saha [67] indicated ion exchange as the adsorption mechanism for removal of MG using alkali-treated fly ash. It was also found that Gibbs free energy (ΔG^0) was spontaneous for all interactions, and the adsorption process exhibited exothermic enthalpy values. Based on the thermodynamic parameters, Gupta et al. (2003) reported that the dye adsorption was spontaneous and endothermic in nature.

8.5   Insights into the mechanisms of adsorption

Adsorption of dyes is basically governed by intra-particle surface diffusion, where pore size distribution and surface area are important.

8.5.1   Implications of instrumentation

Identification and characterization of functional groups on the surface of adsorbents have rewritten the history of adsorption studies. Changes in distribution of functional groups thus revealed the mystery of adsorption by correlating to the influencing parameters such as solution pH, temperature, initial dye concentration, and adsorbent dosage.

Zhang et al. [35] identified some major changes in distribution of functional groups during adsorption of MG onto pine tree root decayed by brown-rot fungi. As lignin was oxidized by enzymes, many polar groups were introduced and the hydroxyl content increased. Also due to demethylation on the aromatic methyl groups of lignin, cellulose and other polysaccharides, the molecular strand could be easily crushed (Table 8.3).

8.5.2   Impact of surface characterization on adsorption capacity

Although BET surface area is small in comparison, the maximum adsorption capacity can be much higher owing to the presence of higher oxygenated surface groups, high average pore diameter, and elevated superficial charge density [63]. Chieng et al. [48] observed that potassium ions were reduced drastically from 31.71% to 0.25% as MG was adsorbed onto breadnut peel by their mutual replacement (Table 8.3). The surface structures of carbon–oxygen (functional groups) are by far the most important structures in influencing the surface characteristics and surface behavior of adsorbent material [72].

The surface charge of the major constituents of the adsorbents plays an important role in the adsorptive removal of cationic dye molecules. The polar functional groups of the adsorbents are involved in the formation of bonds with cationic dyes resulting in reduced adsorption at lower pH apparently due to the competition with hydrogen ions for the adsorption sites [69]. Due to the high percentage of acid minerals in coal (alumina and silica) and many functional groups, at increased pH, the adsorbent surface tends to acquire negative charges, resulting in increased adsorption capacity [70].

8.5.3   Identification of rate-limiting step

Dizge et al. [66] found that intra-particle diffusion and external mass transfer had rate limiting effect on adsorption which was attributed to the relatively simple macro-pore structure of fly ash particle. Similarly, the modeling attempts to identify the rate-limiting step are attributed to merely the surface features of the adsorbent in many other studies irrespective of the inherent chemical characteristics.

8.5.4   Diffusion versus chemisorption

The inherent complexity in exploring the surface features and limited knowledge about their induced chemical interactions make it difficult to estimate the order of reaction rate in bio-sorption process. As advocated by many, it is not proper to assume reaction order initially and try to adjust the model; rather reaction order should be determined based on experimental results. Therefore, these models serve as guidelines for description of process kinetics for the design of operational conditions [44]. At lower concentrations of MG, rate of adsorption governed by external transport whereas at higher concentrations it takes place by internal transport mechanism [32].

Molecular diffusion, in general, consists of essentially three stages, namely, (1) diffusive transport of component from the bulk liquid through sorbent boundary layer, (2) intra-particle diffusion in sorbent pores, and (3) bonding of molecules to active places in sorbent pores. Knowledge of the stage which influences the kinetics of the process is important in designing adsorption systems. The quickest stage, however, is bonding in active places indicating that the other two stages can pose significant rate-limitation under suitable circumstances [44]. Gupta et al. [20] reported that major mass transfer process during adsorption was film diffusion which was controlled by van der Waals and hydrogen bonding.

8.6   Conclusive remarks

Even though adsorption is mostly recommended as a sustainable technology for removal of hazardous chemicals using waste-generated materials, yet adsorption is not free from drawbacks. One of the critical issues is the non-selective nature of the process, competition among sorbates significantly influence the binding capacity of the support material in an unpredictable manner. Additionally, since there is no structural transformation is happening to the dye molecules during adsorption, fate of the regenerating elute as well as disposal of the adsorbent is of prime environmental concern. Among the papers that have been reviewed, some papers suggest intra-particle diffusion as the rate-limiting mechanism. However, some others have identified the chemical mass transfer as the significant mechanism of adsorption. As the current research envisages, simultaneous mineralization or degradation of dyes on the surface of specially prepared reactive support materials could be a viable option to ensure sustainability of these waste management activities. Any large-scale applications based on the adsorption process have to consider these issues discussed above.

References

1. Rafatullah, M.; Sulaiman, O.; Hashim, R.; Ahmad, A. Adsorption of methylene blue on low-cost adsorbents: A review. *J. Hazard. Mater.* **2010**, *177*, 70–80.

2. Dabrowski, A. Adsorption, from theory to practice. *Adv. Colloid Interface Sci.* **2001**, *93*, 135–224.

3. Ahmad, A.; Rafatullah, M.; Danish, M. Removal of Zn(II) and Cd(II) ions from aqueous solutions using treated sawdust of sissoo wood as an adsorbent. *Holz als Roh-und Werkstoff.* **2007**, *65*, 429–436.

4. Rafatullah, M.; Sulaiman, O.; Hashim, R.; Ahmad, A. Adsorption of copper(II), chromium(III), nickel(II), and lead(II) ions from aqueous solutions by meranti sawdust. *J. Hazard. Mater.* **2009**, *170*, 969–977.

5. Dolphen, R.; Sakkayanwong, N.; Thiravetyan, P.; Nakbanpote, W. Adsorption of reactive Red 141 from wastewater onto modified chitin. *J. Hazard. Mater.* **2007**, *145*, 250–255.

6. Kapdan, I. K.; Kargi, F. Simultaneous bio degradation and adsorption of textile dye stuff in an activated sludge unit. *Process BioChem.* **2002**, *37*, 973–981.

7. Ho, Y. S.; McKay, G. The kinetics of sorption of basic dyes from aqueous solutions by sphagnum moss peat. *Can. J. Chem. Eng.* **1998**, *76*, 822–826.

8. Culp, S. J.; Beland, F. A. Malachite green: A toxicological review. *J. Am. Colloid Toxicol.* **1996**, 15, 219–238.

9. Srivastava, S.; Rangana, S.; Roy, D. Toxicological effects of Malachite Green. *Aquat. Toxicol.* **2004**, *66*, 319–329.

10. Metivier-Pignon, H.; Faur-Brasquet, C.; Cloirec, P. L. Adsorption of dyes onto activated carbon cloths: Approach of adsorption mechanisms and coupling of ACC with ultrafiltration to treat coloured wastewaters. *Sep. Purif. Technol.*, **2003**, *31*, 3–11.

11. Ravi, K.; Deebika, B.; Balu, K. Decolourization of aqueous dye solutions by a novel adsorbent: Application of statistical designs and surface plots for the optimization and regression analysis. *J. Hazard. Mater.* **2005**, *B122*, 75–83.

12. McMullan, G.; Meehan, C.; Conneely, A.; Kirby, N.; Robinson, T.; Nigam, P.; Banat, I. M.; Marchant, R.; Smyth, W. F. Microbial decolourisation and degradation of textiles dyes. *Appl. Microbiol. Biotechnol.* **2001**, *56*, 81–87.

13. Pearce, C. I.; Lloyd, J. R.; Guthrie, J. T. The removal of colour from textiles wastewater using whole bacterial cells: A review. *Dyes Pigments.* **2003**, *58*, 179–196.

14. Lee, J. W.; Choi, S. P.; Thiruvenkatachari, R.; Shim, W. G.; Moon, H. Evaluation of the performance of adsorption and coagulation processes for the maximum removal of reactive dyes. *Dyes Pigments.* **2006**, *69*, 196–203.

15. Robinson, T.; McMullan, G.; Marchant, R.; Nigam, P. Remediation of dyes in textiles effluent: A critical review on current treatment technologies with a proposed alternative. *Bioresour. Technol.* **2001**, *77*, 247–255.

16. Banat, I. M.; Nigam, P.; Singh, D.; Marchant, R. Microbial decolourization of textile-dye-containing effluents: A review. *Bioresour. Technol.* **1996**, *58*, 217–227.

17. Selen, M. A. G. U. de Souza; Peruzzo, L. C., de Souza, Antonio, A. U. Numerical study of the adsorption of dyes from textile effluents. *Appl. Math. Modell.* **2008**, *32*, 1711–1718.

18. Zollenger, H. In *Colour Chemistry: Synthesis, Properties and Application of Organic Dyes and Pigment*; **1991**, second revised ed. VHC, Weinheim.

19. Sandra, J. C.; Lonnie, R. B.; Donna, F. K.; Daniel, R. D.; Louis, T.; Frederick, A. B. Toxicity and metabolism of malachite green and leuco malachite green during short-term feeding to Fischer 344 rats and B6C3F1 mice. *Chem. Biol. Interact.* **1999**, *122*, 153–170.

20. Gupta, V. K.; Mittal, A.; Krishnan, L.; Gajbe, V. Adsorption kinetics and column operations for the removal and recovery of malachite green from wastewater using bottom ash. *Sep. Purif. Technol.* **2004**, *40*, 87–96.

21. Kumar, K. V.; Sivanesan, S.; Ramamurthi, V. Adsorption of malachite green onto Pithophora sp., a freshwater alga: Equilibrium and kinetic modeling. *Process Biochem.* **2005**, *40*, 2865–2872.

22. Hoffman, G. L.; Meyer, F. P. In *Parasites of Freshwater Fishes*; **1974**. TFH Publications, Neptune, New Jersey.

23. Ahmad, R.; Kumar, R. Adsorption studies of hazardous malachite green onto treated ginger waste. *J. Environ. Manage.* **2010**, *91*, 1032–1038.

24. Baek, M.-H.; Ijagnemi, C. O.; Se-Jin, O.; Kim, D.-S. Removal of malachite green from aqueous solution using degreased coffee bean. J. Hazard. Mater. **2010**, *176*, 820–828.

25. Dahri, M. K.; Kooh, M. R. R.; Lim, L. B. L. Water remediation using low cost adsorbent walnut shell for removal of malachite green: Equilibrium, kinetics, thermodynamic and regeneration studies. *J. Environ. Chem. Eng.* **2014**, *2*, 1434–1444.

26. Garg, V. K.; Gupta, R.; Yadav, A. B.; Kumar, R. Dye removal from aqueous solution by adsorption on treated sawdust. *Bioresour. Technol.* **2003**, *89*, 121–124.

27. Hameed, B. H.; El-Khaiary, M. I. Batch removal of malachite green from aqueous solutions by adsorption on oil palm trunk fibre: Equilibrium isotherms and kinetic studies. *J. Hazard. Mater.* **2008a**, *154*, 237–244.

28. Hameed, B. H.; El-Khaiary, M. I. Malachite green adsorption by rattan sawdust: Isotherm, kinetic and mechanical modeling. *J. Hazard. Mater.* **2008b**, *159*(2–3), 574–579.

29. Jalil, A. A.; Triwahyono, S.; Yaakob, M. R.; Azmi, Z. Z. A.; Sapawe, N.; Kamarudin, N. H. N.; Setiabudi, H. D.; Jaafar, N. F.; Sidik, S. M.; Adam, S. H.; Hameed, B. H. Utilization of bivalve shell-treated *Zea mays* L. (maize) husk leaf as a low-cost biosorbent for enhanced adsorption of malachite green. *Bioresour. Technol.* **2012**, *120*, 218–224.

30. Khattari, S. D; Singh, M. K. Removal of malachite green from dye wastewater using neem sawdust by adsorption. *J. Hazard. Mater.* **2009**, *167*, 1089–1094.

31. Kumar, V. K. Optimum sorption isotherm by linear and non-linear methods for malachite green onto lemon peel. Dyes Pigments. **2007**, *74*, 595–597.

32. Mittal, A.; Krishnan, L.; Gupta, V. K. Removal and recovery of malachite green from wastewater using an agricultural waste material, de oiled soya. *Sep. Purif. Technol.* **2005**, *43*, 125–133.

33. Saha, P.; Chowdhury, S.; Gupta, S.; Kumar, I.; Kumar, R. Assessment on the removal of malachite green using tamarind fruit shell as biosorbent. *Clean Soil, Air, Water.* **2010**, *38*, 437–445.

34. Sartape, A.; Mandhare, A. M.; Jadhav, V. V.; Raut, P. D.; Anuse, M. A.; Kolekar, S. S. Removal of malachite green dye from aqueous solution with adsorption technique using *Limonia acidissima* (wood apple) shell as low cost adsorbent. Arab. J. Chem. **2017**, *10*(2), S3229–S3238.

35. Zhang, H.; Tang, Y.; Ke, Z.; Su, X.; Cai, D.; Wand, X.; Liu, Y.; Huang, Q.; Yu, Z. Improved adsorptive capacity of pine wood decayed by fungi *Poria cocos* for removal of malachite green from aqueous solution. *Desalination.* **2011**, *274*, 87–104.

36. Hamdaoui, O.; Saoudi, F.; Chiha, M.; Naffrechoux, E. Sorption of malachite green by a novel sorbent, dead leaves of plane tree: Equilibrium and kinetic modeling. *Chem. Eng. J.* **2008**, *143*, 73–84.

37. Pan, X.; Zhang, D. Removal of malachite green from water by *Firmiana simplex* wood fiber. *Electr. J. Biotechnol.* **2009**, *12*(4), 1–10.

38. Wang, H.; Yuan, X.; Zeng, G.; Leng, L.; Peng, X.; Liao, K.; Peng, L.; Xiao, Z. Removal of malachite green dye from wastewater by different organic acid-modified natural adsorbent: Kinetics, equilibriums, mechanisms, practical application, and disposal of dye-loaded adsorbent. *Environ. Sci. Pollut. Res.* **2014**. doi:10.1007/s11356-014-3025-2.

39. Makeswari, M.; Santhi, T. Removal of malachite green dye from aqueous solutions onto microwave assisted zinc chloride chemical activated epicarp of *Ricinus communis*. *J. Water Resour. Protec.* **2013**, *5*, 222–238.

40. Kushwaha, A. K.; Gupta, N.; Chattopadhyaya, M. C. Removal of cationic methylene blue and malachite green dyes from aqueous solution by waste materials of *Daucus carota. J. Saudi Chem. Soc.* **2014**, *18*, 200–207.

41. Sharma, N.; Tiwari, D. P.; Singh, S. K. The efficiency appraisal for removal of Malachite Green by potato peel and neem bark: Isotherm and kinetic studies. *Int. J. Chem. Environ. Eng.* **2014**, *5*(2), 83–88.

42. Gupta, N.; Kushwaha, A. K.; Chattopadhyaya, M. C. Kinetics and thermodynamics of malachite green adsorption on banana pseudo-stem fibers. *J. Chem. Pharm. Res.* **2011**, *3*(1), 284–296.

43. Gupta, N.; Kushwaha, A. K.; Chattopadhyaya, M. C. Adsorption studies of cationic dyes onto Ashoka (*Saraca asoca*) leaf powder. *J. Taiwan Inst. Chem. Eng.* **2012**, *43*, 604–613.

44. Witek-Krowiak, A. Analysis of influence of process conditions on kinetics of malachite green biosorption onto beech sawdust. *Chem. Eng. J.* **2011**, *171*, 976–985.

45. Gupta, N.; Kushwaha, A. K.; Chattopadhyaya, M. C. Application of potato (*Solanum tuberosum*) plant wastes for the removal of methylene blue and malachite green dye from aqueous solution. *Arab. J. Chem.* **2016**, *9*, S707–S716.

46. Kumar, K. V., Porkodi, K. Batch adsorber design for different solution volume/adsorbent mass ratios using the experimental equilibrium data with fixed solution volume/adsorbent mass ratio of malachite green onto orange peel. *Dyes Pigments.* **2007**, *74*, 590–594.

47. Gong, R.; Jin, Y.; Chen, F.; Chen, J.; Liu, Z. Enhanced malachite green removal from aqueous solution by citric acid modified rice straw. *J. Hazard. Mater.* **2006**, *B137*, 865–870.

48. Chieng, H. I.; Lim, L. B. L.; Priyantha, N. Enhancing adsorption capacity of toxic malachite green dye through chemically modified breadnut peel: Equilibrium, thermodynamics, kinetics and regeneration studies. Environ. Technol. **2015**, *36*(1), 86–97.

49. Ozdes, D.; Gundogdu, A.; Duran, C.; Senturk, H.B. Evaluation of adsorption characteristics of Malachite Green onto Almond shell (*Prunus dulcis*). Sep. Sci. Technol. **2010**, *45*, 2076–2085.

50. Singh, J.; Kaur, G. Freundlich, Langmuir adsorption isotherms and kinetics for the removal of malachite green from aqueous solutions using agricultural waste rice straw. Int. J. Environ. Sci. **2013**, *4*(3), 250–258.

51. Kumar, K.V.; Sivanesan, S. Isotherms for malachite green onto rubber wood (*Hevea brasiliensis*) sawdust: Comparison of linear and non-linear methods. *Dyes Pigments*. **2007**, *72*, 124–129.

52. Sonawane, G. H.; Shrivastava, V. S. Kinetics of decolourisation of malachite green from aqueous medium by maize cob (*Zea maize*): An agricultural solid waste. *Desalination*. **2009**, *247*, 430–441.

53. Franca, A. S.; Oliveira, L. S.; Saldanha, S. A.; Santos, P. I. A.; Salum, S. S. Malachite green adsorption by mango (*Mangifera indica* L.) seed husks: Kinetic, equilibrium and thermodynamic studies. *Desalination Water Treat*. **2010**, *19*, 241–248.

54. Santhi, T.; Manonmani, S. Malachite green removal from aqueous solution by the peel of *Cucumis sativa* fruit. *Clean Soil, Air, Water*. **2011**, *39*(2), 162–170.

55. Wang, X. S.; Zhou, Y.; Jiang, Y.; Sun, C. The removal of basic dyes from aqueous solutions using agricultural by-products. *J. Hazard. Mater.* **2008**, *157*, 374–385.

56. Papinutti, L.; Mouso, N.; Forchiassin, F. Removal and degradation of the fungicide dye malachite green from aqueous solution using the system wheat bran-*Fomes scrlerodermeus*. *Enzyme Microb. Technol.* **2006**, *39*, 848–853.

57. Dahri, M. K.; Kooh, M. R. R.; Lim, L. B. L. Application of *Casuarina equisetifolia* needle for the removal of methylene blue and malachite green dyes from aqueous solution. Alexa. Eng. J. **2015**, *54*, 1253–1263.

58. Srivastava, R.; Rupainwar, C. D. A comparative evaluation for adsorption of dye on neem bark and mango bark powder. *Ind. J. Chem. Technol.* **2011**, *18*, 67–75.

59. Chowdhury, S.; Mishra, R.; Saha, P.; Kushwaha, P. Adsorption thermodynamics, kinetics and isosteric heat of adsorption of malachite green onto chemically modified rice husk. *Desalination*. **2011**, *265*, 159–168.

60. Forgacs, E.; Cserhati, T.; Oros, G. Removal of synthetic dyes from wastewaters: A review. *Environ. Int.* **2004**, *30*, 953–971.

61. Bharathi, K. S.; Ramesh, S. T. Removal of dyes using agricultural waste as low-cost adsorbents: A review. *Appl. Water Sci.* **2013**, *3*(4), 773–790.

62. Garg, V. K.; Gupta, R.; Kumar, R.; Gupta, R.K. Adsorption of chromium from aqueous solution on treated sawdust. *Bioresour. Technol.* **2004**, *92*, 79–81.

63. Mendez. A.; Barriga. S; Fidalgo, J. M.; Gasco, G. Adsorbent materials from paper industry waste materials and their use in Cu(II) removal from water. *J. Hazard. Mater.* **2010**, *165*(1–3), 736–743.

64. Gurkan, E. H.; Semra, C. Using waste foundry sand for the removal of malachite green dye from aqueous solutions: Kinetic and equilibrium studies. *Environ. Eng. Manage. J.* **2018**, 17(1), 123–133.

65. Dincer, A. R.; Gunes, Y.; Karakaya, N. Coal based bottom ash (CBBA) waste material as adsorbent for removal of textile dyestuffs from aqueous solution. *J. Hazard. Mater.* **2007**, *141*(3), 529–535.

66. Dizge, N.; Aydiner, C.; Demirbas, E.; Kobya, M.; Kara, S. Adsorption of reactive dyes from aqueous solutions by fly ash: kinetic and equilibrium studies. *J. Hazard. Mater.* **2008**, *150*(3), 737–746.

67. Chowdhury, S.; Saha, P. Adsorption thermodynamics and kinetics of malachite green onto Ca(OH)$_2$ treated fly ash. *J. Environ. Eng.* **2011**, *137*(5), 388–397.

68. Khan, T. A.; Ali, I.; Singh, V. V.; Sharma, S. Utilisation of fly ash as low-cost adsorbent for the removal of methylene blue, malachite green and rhodamine B dyes from textile wastewater. *J. Environ. Protect.* **2009**, 3(1), 11–22.

69. Khan, T. A.; Singh, V. V.; Kumar, D. Removal of some basic dyes from artificial textile wastewater by adsorption on Akash Kinari coal. *J. Sci. Ind. Res.* **2004**, 63(4), 355–364.

70. Kooh, M. R. R.; Dahri, K. M.; Lim, L. B. L. Jackfruit seed as low-cost adsorbent for removal of malachite green: Artificial neural network and random forest approaches. *Environ. Earth Sci.* **2018**, 77, 434.

71. Krishna Murthy, T. P.; Gowrishankar, B. S.; Prabha, M. N. C.; Kruthi, M.; Hari Krishna, R. Studies on batch adsorptive removal of malachite green from synthetic wastewater using acid treated coffee husk: Equilibrium, kinetics and thermodynamic studies. *Microchem. J.* **2019**. doi.org/10.1016/j.microc.2018.12.067\.

72. Mall, I. D.; Srivastava, V. C.; Agarwal, N. K.; Mishra, I. M. Adsorptive removal of malachite green dye from aqueous solution by bagasse fly ash and activated carbon-kinetic study and equilibrium isotherm analyses. *Colloids Surf. A.* **2005**, *264*(1–3), 17–28.

73. Mane, V. S.; Mall, I. D.; Srivastava, V. C. Kinetic and equilibrium isotherm studies for the adsorptive removal of brilliant green dye from aqueous solution by rice husk ash. *J. Environ. Manage.* **2007**, *84*(4), 390–400.

74. Mohammad, M.; Maitra, S.; Dutta, B. K. Comparison of activated carbon and physic seed hull for the removal of malachite green dye from aqueous solution. *Water Air Soil Pollut.* **2018**, *229*, 45.

75. Odoemelam, S. A.; Emeh, N. U.; Eddy, N. O. Experimental and computational chemistry studies on the removal of methylene blue and malachite green dyes from aqueous solution by neem (*Azadirachta indica*) leaves. *J. Taibah Univ. Sci.* **2018**, *12*(8), 255–265.

76. Wang, S.; Li, H. Dye adsorption on unburnt carbon: Kinetics and equilibrium. *J. Hazard. Mater.* **2005**, *126*(1–3), 71–77.

77. Gupta, V. K.; Srivastava, S. K.; Mohan, D. Equilibrium uptake, sorption dynamics, process optimization and column operations for the removal and recovery of malachite green from wastewater using activated carbon and activated slag. *Ind. Eng. Chem. Res.* **1997**, *36*(6), 2207–2218.

78. Zhang, L.; Zhang, H.; Tian, Y. Removal of malachite green and crystal violet cationic dyes from aqueous solution using activated carbon process red mud. *Appl. Clay Sci.* **2014**, *93*, 85–93.

79. Khan, T. A.; Rahman, R.; Ali, I.; Khan, E. A.; Mukhlif, A. A. Removal of malachite green from aqueous solution using waste pea shells as low cost adsorbent – adsorption isotherms and dynamics. *Toxicol. Environ. Chem.* **2014**, *96(4)*, 569–578. https://doi.org/10.1080/02772248.2014.969268.

80. Kumar, K.V. Optimum sorption isotherm by linear and non-linear methods for malachite green onto lemon peel. *Dye. Pigments* **2007**, *74(3),* 595–597.

81. Gupta, V. K.; Gupta, R.; Yadav, R. B.; Kumar, R. Dye removal from aqueous solutions by adsorption on treated sawdust. *Bioresour. Technol.* **2003**, *89,* 121–124.

82. Mittal, A.; Mittal, J.; Kurup, L.; Singh, A.K. Process development for removal and recovery of hazardous dye erythrosine from wastewater by waste materials – bottom ash and de-oiled soya as adsorbents. *J. Hazard. Mater.* **2002**, *138,* 95–105.